UNIVERSITY OF BATH • SCIENCE 16-19

Other titles in the Project

Physics Robert Hutchings
Telecommunications John Allen
Medical Physics Martin Hollins
Nuclear Physics David Sang

Biology Martin Rowland
Applied Genetics Geoff Hayward
Applied Ecology Geoff Hayward
Micro-organisms and Biotechnology Jane Taylor
Biochemistry and Molecular Biology Moira Sheehan

Chemistry Ken Gadd and Steve Gurr

UNIVERSITY OF BATH • SCIENCE 16-19

Project Director: J. J. Thompson, CBE

ENERGY

DAVID SANG
ROBERT HUTCHINGS

Thomas Nelson and Sons Ltd
Nelson House   Mayfield Road
Walton-on-Thames   Surrey
KT12 5PL   UK

51 York Place
Edinburgh
EH1 3JD   UK

Thomas Nelson (Hong Kong) Ltd
Toppan Building 10/F
22A Westlands Road
Quarry Bay   Hong Kong

Thomas Nelson Australia
102 Dodds Street
South Melbourne
Victoria 3205   Australia

Nelson Canada
1120 Birchmount Road
Scarborough   Ontario
M1K 5G4   Canada

First published by Macmillan Education Ltd 1991
ISBN 0-333-53109-4

This edition published by Thomas Nelson and Sons Ltd 1992

ISBN 0-17-448190-X
NPN 9 8 7 6 5 4 3

Printed in Hong Kong.

Contents

The Project: an introduction

The **University of Bath · Science 16–19 Project**, grew out of a reappraisal of how far sixth form science had travelled during a period of unprecedented curriculum reform and an attempt to evaluate future development. Changes were occurring both within the constitution of 16–19 syllabuses themselves and as a result of external pressures from 16+ and below: syllabus redefinition (starting with the common cores), the introduction of AS-level and its academic recognition, the originally optimistic outcome to the Higginson enquiry; new emphasis on skills and processes, and the balance of continuous and final assessment at GCSE level.

This activity offered fertile ground for the School of Education at the University of Bath to join forces with a team of science teachers, drawn from a wide spectrum of educational experience, to create a flexible curriculum model and then develop resources to fit it. This group addressed the task of satisfying these requirements:

- the new syllabus and examination demands of A- and AS-level courses;
- the provision of materials suitable for both the core and options parts of syllabuses;
- the striking of an appropriate balance of opportunities for students to acquire knowledge and understanding, develop skills and concepts, and to appreciate the applications and implications of science;
- the encouragement of a degree of independent learning through highly interactive texts;
- the satisfaction of the needs of a wide ability range of students at this level.

Some of these objectives were easier to achieve than others. Relationships to still evolving syllabuses demand the most rigorous analysis and a sense of vision – and optimism – regarding their eventual destination. Original assumptions about AS-level, for example, as a distinct though complementary sibling to A-level, needed to be revised.

The Project, though, always regarded itself as more than a provider of materials, important as this is, and concerned itself equally with the process of provision – how material can best be written and shaped to meet the requirements of the educational market-place. This aim found expression in two principal forms: the idea of secondment at the University and the extensive trialling of early material in schools and colleges.

Most authors enjoyed a period of secondment from teaching, which not only allowed them to reflect and write more strategically (and, particularly so, in a supportive academic environment) but, equally, to engage with each other in wrestling with the issues in question.

The Project saw in the trialling a crucial test for the acceptance of its ideas and their execution. Over one hundred institutions and one thousand students participated, and responses were invited from teachers and pupils alike. The reactions generally confirmed the soundness of the model and allowed for more scrupulous textual housekeeping, as details of confusion, ambiguity or plain misunderstanding were revised and reordered.

The test of all teaching must be in the quality of the learning, and the proof of these resources will be in the understanding and ease of accessibility which they generate. The Project, ultimately, is both a collection of materials and a message of faith in the science curriculum of the future.

J.J. Thompson
January 1990

How to use this book

In the last decades of the twentieth century, ideas about energy have become of increasing public interest and concern. Problems of energy consumption and conservation are at the heart of many of the stories in our daily newspapers. Fuel shortages, alternative technologies, worries about radiation – all of these are subjects of discussion.

To make sense of these discussions, it is important to have an understanding of what Science has to tell us about energy. In your earlier studies of Science, you will have looked at many aspects of energy. We have written this book to help you to extend your understanding of fundamental principles, and to give you some insight into more advanced ideas about energy.

The book is divided into three themes. The first deals with energy as an important human resource. How do we get the energy we use? How do we use it? What are the technologies involved? What are future trends in energy consumption likely to be? Where possible, we have asked you to reflect on some of the problems which result from our use of energy. These questions may be ultimately political, and we do not expect you to give a 'correct' answer; however, you should be able to see how scientific knowledge can influence political judgements.

The second theme is rather harder, in that it deals with the laws of thermodynamics. The First Law is familiar as the Principle of Conservation of Energy; the Second Law has important consequences for the extent to which we can benefit from available energy resources. It has been suggested that the Second Law is the scientific equivalent of the works of Shakespeare, and that it should be a subject of compulsory study for all politicians!

The third theme is about transport. Our transport systems use highly-developed energy-based technologies. This theme studies the structures and forces involved.

There are a number of activities throughout the text for you to carry out in addition to reading, which on its own may be too passive to promote effective thinking and learning. Questions in the text have two purposes: to help you to check and consolidate your understanding as you proceed, and to encourage you to think ahead, to work out the next steps in the argument. The assignments and investigations will help you to practise using your knowledge. At the end of each theme are some examination questions, to help you to assess whether you have reached an appropriate standard in your studying.

We have assumed that you have someone to guide your learning, and to help you out if you get really stuck. We have tried to write this book so that it will support your learning, but if in doubt, ask your teacher.

For most people, it is easiest to study in collaboration with a partner. You can set each other targets, help each other over the difficult bits, try out the assignments together, and generally make your learning more enjoyable.

Energy issues will continue to be important throughout your lifetime. We hope that a greater understanding of the science behind these questions will help you as a citizen of the world; perhaps as a future scientist or engineer, you will be able to play a part in solving some of them.

Learning objectives

These are given at the beginning of each chapter and they outline what you should gain from the chapter. They are statements of attainment and often link closely to statements in a course syllabus. Learning objectives can help you make notes for revision, especially if used in conjunction with the summaries at the end of the chapter, as well as for checking progress.

Questions

In-text questions occur at points when you should consolidate what you have just learned, or prepare for what is to follow by thinking along the lines required by the question. Some questions can, therefore, be answered from the material covered in the previous section, others may require additional thought or information. Answers to numerical questions are at the end of the book.

Examination questions

At the end of each theme is a group of examination questions relating to the topics covered in the theme. These can be used to help consolidate understanding of the theme or for revision at the end of the course.

Assignments and Investigations

Where you are asked to think about a particular idea, or to develop an idea further, you will find text and questions presented together as an assignment. You will find guidelines to help you at the end of the book.

We have not given full details of experimental methods, except where these are necessary for reasons of safety. You should be able to decide for yourself what quantities need to be measured, and how to measure them.

Summaries

Each chapter ends with a brief summary of its content. These summaries, together with the learning objectives, should give you a clear overview of the subject, and allow you to check your own progress.

Other resources

We have assumed that you will have access to various useful books and software; in particular, you will need to have a data book which lists the properties of different radioactive nuclides, and perhaps a chart of radionuclides. We have listed some suitable books and some useful addresses in Appendix C.

Acknowledgements

The author and publishers wish to thank the following who have kindly given permission for the use of copyright material:

Friends of the Earth for an extract from *Energy Without End* by Michael Flood. University of Cambridge Local Examinations Syndicate and University of London School Examinations Board for questions from past examination papers.

The authors and publishers wish to acknowledge, with thanks, the following photographic sources:

Barnaby's Picture Library *pp 49, 94;* BP Solar *p 5;* J. Allan Cash *pp 11, 109, 110;* Coventry Polytechnic Energy Systems Group *p 13 lower;* Eling Tide Mill Trust *p 9;* Environmental Picture Library *p 18 right;* ETSU *pp 13 upper, 14;* European Space Agency *p 71;* Ford Photographic *pp 81, 95;* Hants County Architects (Joe Low) *p 4;* JET Laboratory *p 24;* LEC Refrigeration *p 59 upper;* Lennox Industries *p 80;* Marlec Engineering Co *p 16;* Museum of Childhood, Edinburgh *p 59 lower;* National Grid Co *p 36;* National Motor Museum, Beaulieu *p 89 upper;* Dave Neal *p 32;* Panos Pictures *pp 1 upper, 18 upper left;* Press Association *p 1 lower;* Powergen *p 76;* Rolls Royce plc *pp 69, 83;* Trustees of the Science Museum *p 6;* Scottish Hydro-Electric plc *pp 7, 15;* Shell *pp 89 lower, 96;* Telecom Technology Showcase *p 2;* Topham Picture Source *p 113;* Weald & Downland Open Air Museum *p 18 lower left.*

Every effort has been made to trace all the copyright holders, but if any have been inadvertently overlooked the publishers will be pleased to make the necessary arrangement at the first opportunity.

Theme 1

ENERGY FOR PEOPLE

In the course of your studies, you will no doubt have been told many times that the word 'energy' has a special meaning in science, different from its many meanings in everyday life. In this theme, we will be exploring the scientific view of energy. However, that does not mean that we will not be thinking about the real, everyday world.

Many of the problems faced by people are to do with energy. Perhaps some of us use too much; certainly some of us do not get enough. These problems lie behind many of the stories in today's news. Science has a lot to say about these problems, though it cannot pretend to have all the solutions.

In this theme, we look at the great variety of sources of energy available to us, and how that energy can be distributed. Finally, we look at the patterns of energy resources and consumption, both historically and geographically.

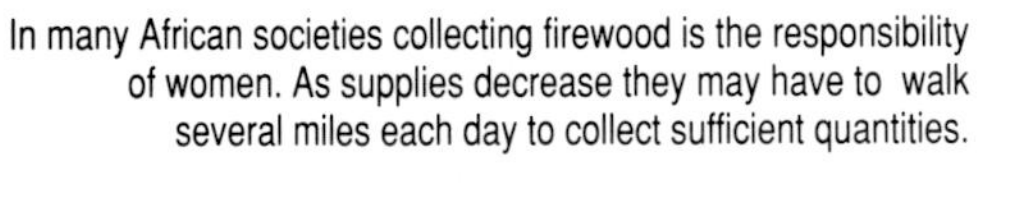

In many African societies collecting firewood is the responsibility of women. As supplies decrease they may have to walk several miles each day to collect sufficient quantities.

British motorists queuing for petrol during a petrol shortage in the winter of 1973/4.

Chapter 1

THE PHYSICS OF ENERGY SUPPLIES

People use a great range of energy sources. Some of these, such as wood for burning, are ancient technologies; others are very recent – think about solar cells and nuclear power. In this chapter, we will consider a variety of the most important energy technologies, and identify the physical principles which underlie them.

LEARNING OBJECTIVES

After studying this chapter you should be able to:

1. explain the physical principles of a variety of energy supplies, including solar energy, wind and water power, biomass energy, fossil and nuclear fuels and geothermal energy;
2. distinguish between energy sources having their origins in recent and ancient solar energy and in the formation of the Earth;
3. distinguish between renewable and non-renewable energy sources.

1.1 ENERGY FROM THE SUN

Most of the energy we use comes, directly or indirectly, from the Sun. The Sun bathes the Earth with radiation – infrared, visible light and ultraviolet. Of course, without the Sun's rays, life on Earth would not have developed.

The solar constant

Satellites have been used to measure the Sun's radiation above the Earth's atmosphere; indeed, many satellites rely on solar cells for their energy supply – Fig 1.1. The intensity of the Sun's radiation is described by the solar constant. The amount of solar energy falling on one square metre at right angles to the Sun's rays in one second is

$$\text{solar constant} = 1373\ \text{W m}^{-2}$$

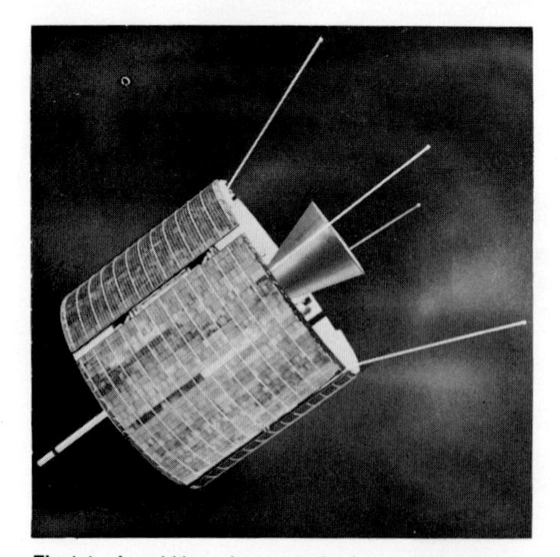

Fig 1.1 An orbiting telecommunications satellite; its batteries are continuously recharged by panels of solar cells.

This quantity is called a constant, although it can vary by up to 30 W m^{-2}, depending on the activity of the Sun.

The Earth reflects some radiation, about 30% of the total. Consequently, the amount of radiation available to the Earth is reduced to about 1 kW m^{-2}. The intensity of the radiation at the Earth's surface is further reduced by absorption by the atmosphere.

In northern European countries, we each typically consume energy at an average rate of about 2 kW. If we could capture the Sun's energy efficiently, only a few square metres per person would satisfy our needs. However, if we are to consider making use of this energy, there are other factors we must take into account.

The **solar constant** describes the power falling on 1 m^2 of surface at right angles to the Sun's rays. The power falling on 1 m^2 of horizontal surface is greatest when the Sun is directly overhead. It varies with the time of day, and with the latitude. It depends on cloud cover, atmospheric pollution and the height above sea level. Fig 1.2 shows the variation of average solar radiation within the UK.

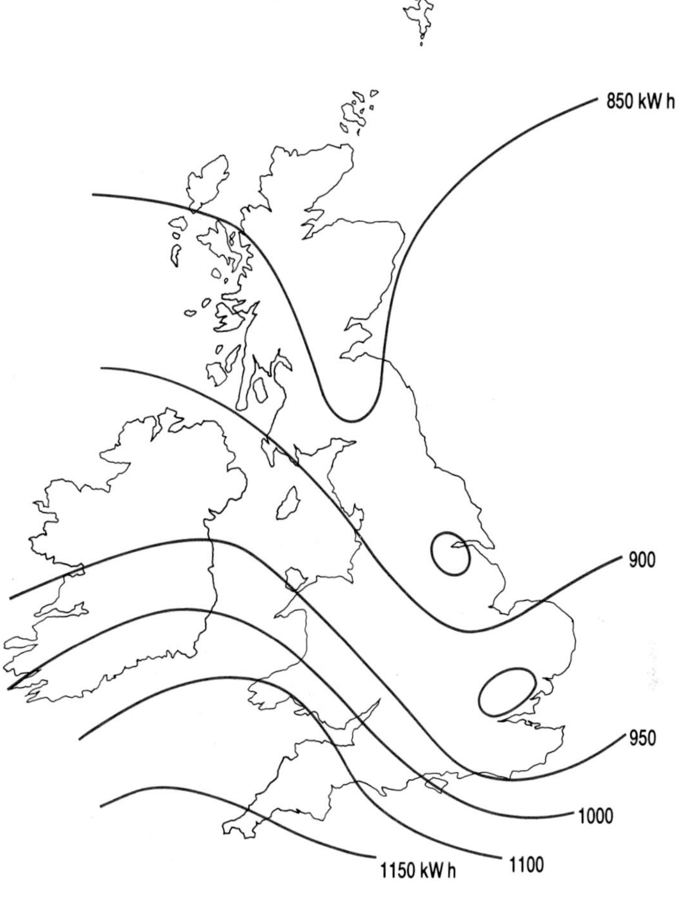

Fig 1.2 The average solar radiation available within the United Kingdom per square metre per year.

All of these factors must be taken into account when considering the possibility of making use of the Sun's energy.

The inverse square law

The Sun's rays spread out as they travel out into space; their intensity decreases as they spread out. The energy falling on each square metre of area decreases as the square of the distance – Fig 1.3. Mathematically, this is written as

$$\text{intensity} \propto 1/r^2$$

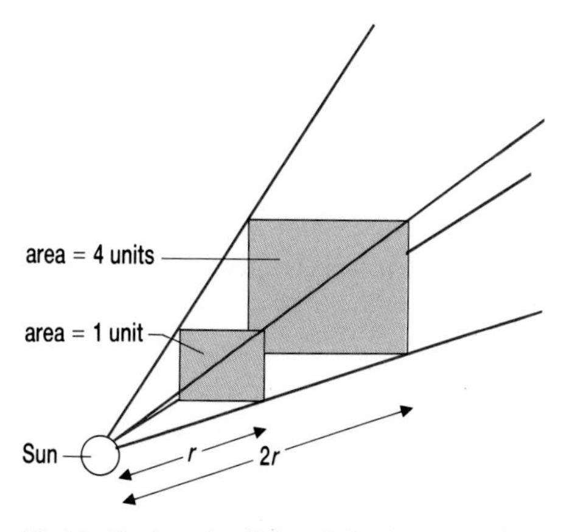

Fig 1.3 The intensity of light radiation decreases with distance according to the inverse square law.

Of course, this law applies to the intensity of light from any point source.

We can use the inverse square law to estimate the variation of the solar constant as the Earth orbits the Sun – Fig 1.4. The Earth's orbit is elliptical; our distance from the Sun varies between 147×10^6 km in December to 152×10^6 km in June. If the average solar constant is 1373 W m^{-2}, what are its maximum and minimum values?

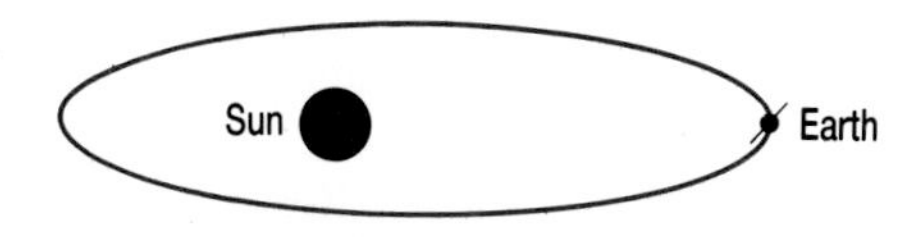

Fig 1.4 The elliptical orbit of the earth about the sun.

The average distance of the Earth from the Sun is 150×10^6 km. In December, when we are closer to the Sun, the solar constant will be greater than its average value:

$$\begin{aligned}\text{solar constant} &= 1373 \times (152/150)^2 \\ &= 1410\ \text{W m}^{-2}\end{aligned}$$

In June, the solar constant is less:

$$\begin{aligned}\text{solar constant} &= 1373 \times (147/150)^2 \\ &= 1319\ \text{W m}^{-2}\end{aligned}$$

Capturing the Sun's energy

One of the major uses of energy in temperate regions of the Earth is space-heating – providing warmth in buildings. The energy falling on the outside of a typical house is several times the energy requirements for heating the inside of the house. By careful design, much of the Sun's radiation falling on a building can be captured. This is known as **passive solar heating**.

How can this energy be captured? All houses absorb some of the Sun's radiation. Windows are important; visible light is transmitted through the glass, and is absorbed, warming up the house. Little heat escapes through the window, since the glass is opaque to infrared rays. This is the principle on which greenhouses and conservatories work.

A school, well-designed from the point of view of passive solar heating, is shown in Fig 1.5. Large, south-facing windows are double-glazed to minimise heat losses by conduction. North-facing windows are small, and may be triple-glazed.

Fig 1.5 This school in Netley, Hampshire is an example of passive solar architecture – it is designed to maximise the capture of the sun's radiation, and to retain heat.

A Trombe wall is designed to maximise energy capture. It consists of a black wall, facing the Sun, and covered with glass. The black wall is a good absorber and reservoir of heat; air circulates between wall and glass, and is ducted into the building. Trombe walls are being tested for use in low-energy housing developments. (A solar panel is a smaller version; note that solar panels heat water, they do not produce electricity.) These are examples of **active solar heating**.

Electricity from the Sun

Solar cells (also known as photocells) use the **photovoltaic effect**. When light falls on certain semiconductor materials, a voltage is generated. Some of the electrons within the material are only weakly bonded, and they may be released if they can capture the energy of sunlight falling on the cell. Fig 1.6 shows the construction of a silicon solar cell.

Solar cells provide a small voltage, usually less than one volt. They are connected in series to provide higher voltages, and in parallel to provide high currents. Unfortunately, they are relatively poor converters of radiation energy to electrical energy, with efficiencies usually less than 20%. This inefficiency arises as follows: each photovoltaic material has an optimum frequency at which it operates. Light of lower frequency cannot release electrons, and is absorbed as heat. Higher frequency light is too energetic, and the excess energy also becomes heat. Research is concentrating on producing materials which have the right characteristics to enable them to waste as little of the Sun's radiation as possible.

Fig 1.7 Solar cells provide electricity to operate this radio station in a remote region of Peru.

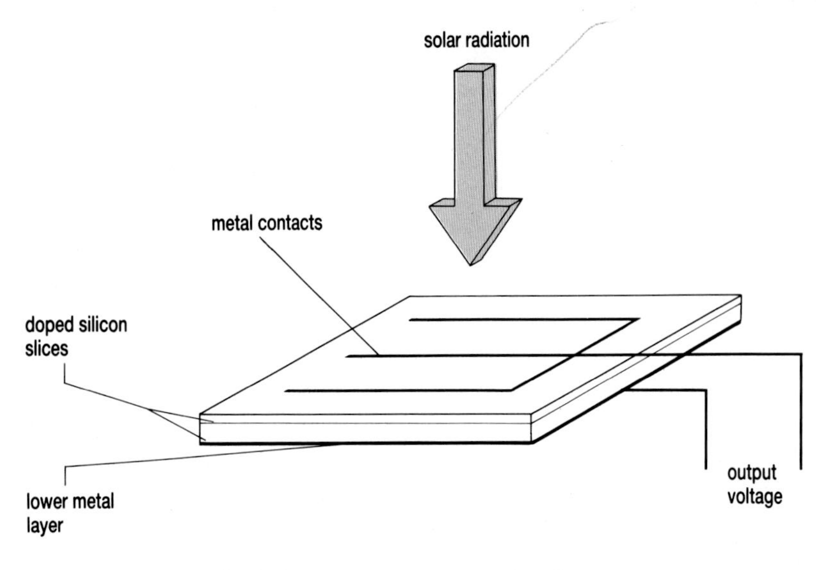

Fig 1.6 The construction of a silicon solar cell.

As the cost of solar cells decreases, they are used in increasing numbers in a wide variety of situations, from calculators to navigation beacons in remote locations. Although they are at present expensive compared with other ways of generating electricity, they are robust and long-lived, and have a number of specialised uses where cost is of secondary importance – see Fig 1.7.

ASSIGNMENT

Making use of solar cells

The first solar cells, which were very expensive, were developed for use as energy sources for satellites. Since then, as Fig 1.8 shows, the cost has dropped, and solar cells are produced in much greater numbers.

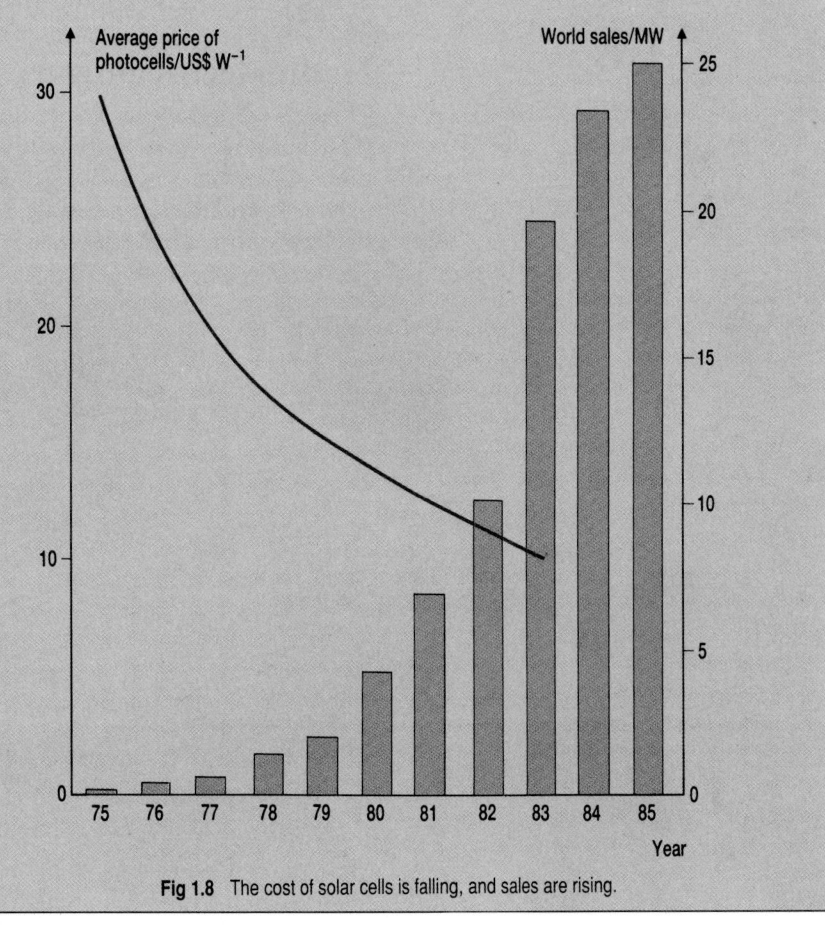

Fig 1.8 The cost of solar cells is falling, and sales are rising.

In answering the questions below, you will have to draw on the information and ideas discussed in this section of the text.

1. Why do you think solar cells found their first application in space technology?
2. A particular satellite consumes electrical energy at an average rate of 20 W. Estimate the area of the solar panel which could provide this power for the satellite in low Earth orbit. (Assume the panel has an efficiency of 10% in converting solar energy to electrical energy.)
3. If the satellite is to travel to Mars, how big would the panel need to be in order to supply the same power? (Radius of Earth's orbit = 150×10^6 km, radius of Mars' orbit = 230×10^6 km.)
4. It has been suggested that, if the cost of solar cells drops to \$1 per peak watt, they could compete effectively with other methods of generating electricity on a large scale. Try to make a realistic estimate of the ground area required for a fairly modest 50 MW solar electric power station, capable of supplying the needs of a town of population 25 000. In making your estimate, take account of the latitude of the town – typically 55° N for the UK. The Sun does not shine 24 hours a day. How would you take account of this in your estimate, and in the design of such a power station?

1.2 WATER POWER

Before the advent of steam, most of our requirements for mechanical power were supplied by water and wind power, and by the exertions of men and animals. The harnessing of the power of running water and moving air is an ancient technology. In this section, we will look at the physics of hydroelectric, tidal, wave and wind power systems.

Hydroelectric power

There is evidence that cornmills driven by water wheels have existed for two thousand years, both in Europe and in China – see Fig 1.9. With the development of electricity generation in the 19th century, it was a simple extension to attach a generator to the rotating kingpin of such a mill, and hence produce a hydroelectric power (HEP or 'hydro') scheme.

Fig 1.9 This water mill of about 1590 shows the way energy from the stream is used to turn giant milling stones to grind corn. Cogs were used to transfer the turning action of the wheel to the two stones at the top of the staging.

Hydroelectric power uses the gravitational potential energy of water in high rivers or lakes as its source of energy. This energy originally comes from the Sun's radiation, which evaporates water from the Earth's surface –

about one quarter of the Sun's energy is used up in this way. The water returns to the Earth's surface as precipitation, and if it falls in the right place, its energy may be useful to us.

In a typical HEP scheme, water is stored behind a dam and released at a controlled rate to flow through a turbine which turns a generator. The dam and power station need not be close together. Indeed, since it is desirable to have a large head of water, it is often the case that they are some distance apart. Fig 1.10 shows two views of the Loch Sloy scheme, Britain's most powerful conventional HEP scheme. The dam is in the mountains, 285 m above sea level. Water runs down huge pipes to the power station on the banks of Loch Lomond, 277 m below.

Fig 1.10 **(a)** The Loch Sloy dam in the Scottish Highlands; **(b)** water from the dam is piped to this 130 MW power station on the banks of Loch Lomond.

Such a 'high-head' system may use a Pelton wheel, Fig 1.11(a), to capture the kinetic energy of the moving water. The water is squirted through a nozzle at a rotating wheel of double cups. The impulse of the water keeps the wheel turning.

Systems where the water falls through a shorter distance use one of a variety of types of turbine which are completely immersed in the flow of water. Slow-moving water enters the turbine and accelerates as it leaves. (This is the principle of a garden lawn sprinkler.) The turbine is turned by the reaction of the water on the blades. An example, a Francis turbine, is shown in Fig 1.11(b).

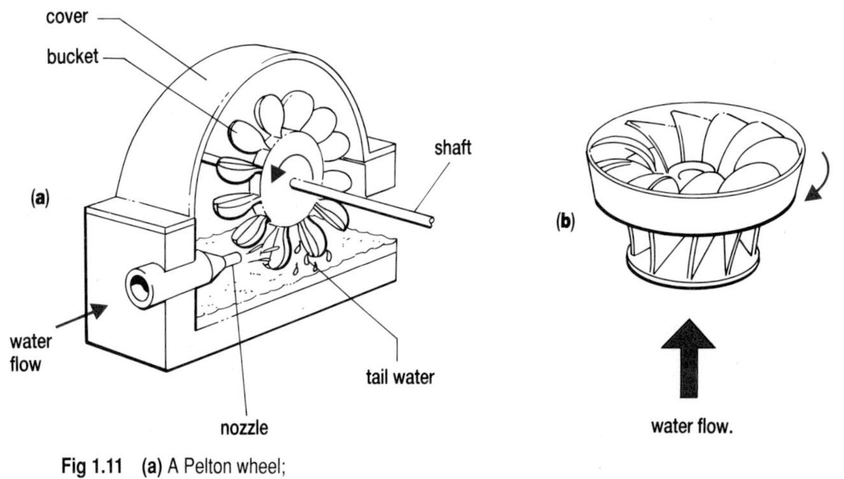

Fig 1.11 **(a)** A Pelton wheel;
(b) a Francis turbine; two types of turbine wheel used in hydroelectric power stations.

Such turbines have been developed considerably over the last hundred years. In order to maximise the efficiency of an HEP scheme, it is necessary to extract as much of the water's kinetic energy as possible. This means that the water leaving the turbine must be moving as slowly as possible. Turbine efficiency is usually between 80 and 90%.

Many countries rely heavily on hydroelectric power as a major source of electricity. In northern Scotland, a large proportion of the available sites have been used, and now 99% of homes in the Highlands and Islands are connected to the grid. Brazil obtains 80% of its electricity from HEP schemes. The giant Itaipu scheme on the river Parana is the biggest in the world, supplying 12 600 MW of power for both Brazil and Paraguay.

Calculating efficiency

To determine the efficiency of any energy conversion scheme, we need to compare the available energy to the useful energy produced, since

$$\text{efficiency} = \frac{\text{energy out}}{\text{energy in}} \times 100\%$$

$$= \frac{\text{power out}}{\text{power in}} \times 100\%$$

For an HEP scheme, the available energy is the gravitational potential energy of the water behind the dam. This is determined by the mass of water stored, and the height of the head of water, that is, the height the water falls between the dam and the turbine.

If we assume that, when a mass m of water falls through a height h, its potential energy E_p is converted entirely to kinetic energy E_k, we can write

$$mgh = \tfrac{1}{2}mv^2$$

and hence the speed v of the water at the turbine is given by

$$v = \sqrt{2gh}$$

The available power depends on the rate at which water passes through the turbine. If x kg s^{-1} pass through, the available power is xgh W. This can be compared with the power output to find the efficiency.

ASSIGNMENT

The Foyers hydroelectric scheme

To obtain a realistic picture of the efficiency of a large HEP scheme, consider as an example the 300 MW station at Foyers on the banks of Loch Ness, Fig 1.12. Table 1.1 gives the relevant data.

Table 1.1 Data for the Foyers HEP scheme

Gross head	179 m
Rate of flow	205×10^3 kg s^{-1}
Power output	300 MW

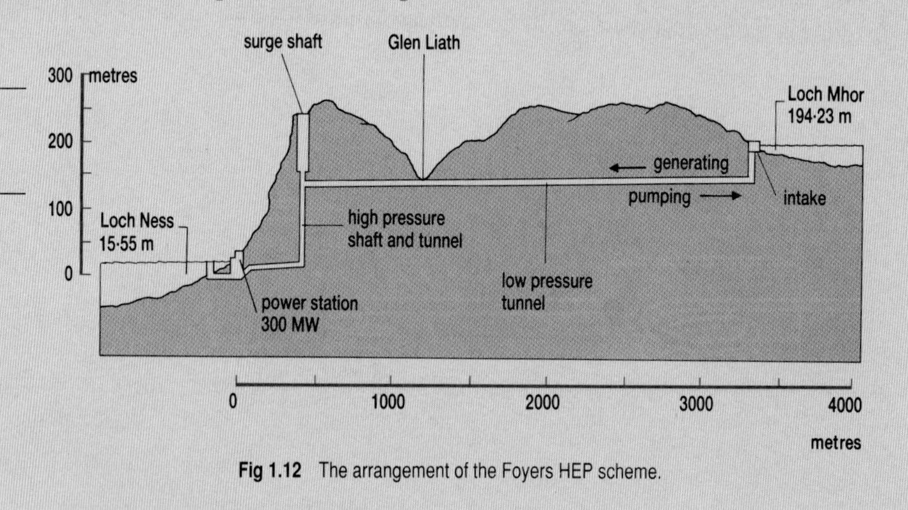

Fig 1.12 The arrangement of the Foyers HEP scheme.

1. Calculate the gravitational potential energy of each kilogram of water stored behind the dam.
2. Estimate the speed of the water arriving at the turbines. Will your answer be an over- or under-estimate?

3. Calculate the energy available to the turbines each second, and hence the efficiency of the HEP station as a whole.
4. Think about ways in which the available energy may be wasted. It may be that the water leaving the turbine is warmer than the water leaving the dam. Calculate the temperature rise of the water on the assumption that all the wasted energy is used up in heating the water. (The specific heat capacity of water is 4200 J kg^{-1} K^{-1}.)
5. Suggest three other ways in which the energy of the water stored behind the dam may fail to be converted wholly to electrical energy.

Tidal power

The energy of the oceans' tides has been used for centuries. Fig 1.13 shows an 18th-century tide mill in Hampshire, on a site where tidal power has been used for milling for nearly six hundred years. A pond behind such a mill filled with water as the tide came in, and the water was released at low tide, giving about eight hours of milling each day.

Fig 1.13 The Eling Tidal Mill at Totton, Hampshire. The mill race can be seen near the bottom of the steps. There were once over a 100 tidal mills around the British coastline.

Tidal energy arises from the gravitational pull of the Moon and the Sun on the Earth's oceans. The water is attracted so that its level in mid-ocean rises by rather less than a metre. Fig 1.14 shows an exaggerated view of this. As the Earth rotates, friction between the sea and the seabed and coastlines extracts energy. The entire heating effect of tidal friction over the whole globe is of the order of 10^{12} W, comparable to the world's total power consumption. However, only a tiny fraction of this energy could conceivably be harnessed for our use.

For tidal energy to be useful, special circumstances must exist. The **tidal range** – the difference in heights between high and low tides – must be large. Tides tend to be magnified in particular locations, for example, where the sea enters a narrowing estuary. Such a site may also be useful, since an estuary may be relatively easily dammed to provide a pool to hold the water.

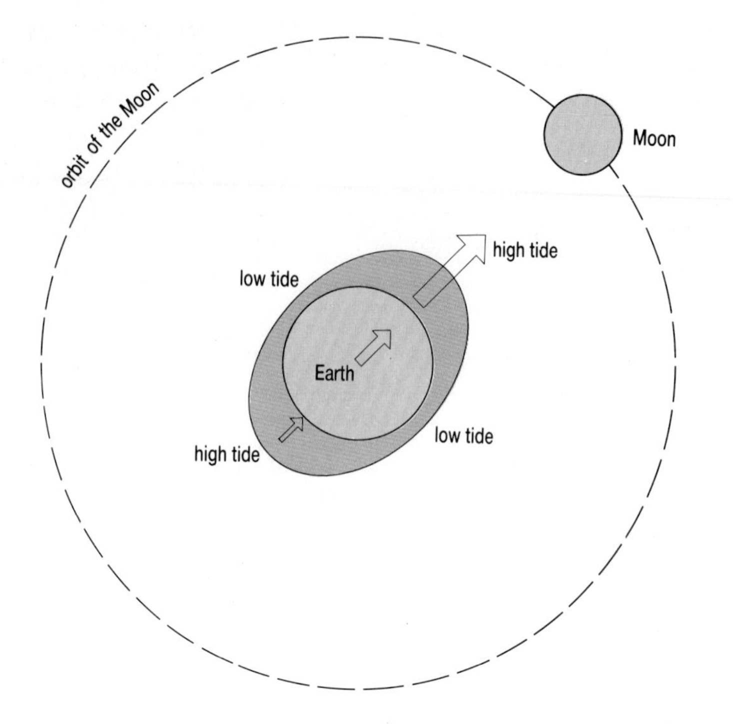

Fig 1.14 Tides are caused by the gravitational attraction of the Moon and, to a lesser extent, the Sun.

The Severn Estuary in south-west England is famous for its tidal bore, and has been the subject of several technical enquiries into its suitability for tidal power. The average tidal range is 9.8 m. A variety of schemes has been suggested (Fig 1.15): single barrages, and more complex two-basin schemes, in which water from a basin filled at high tide flows into a second basin previously emptied at low tide.

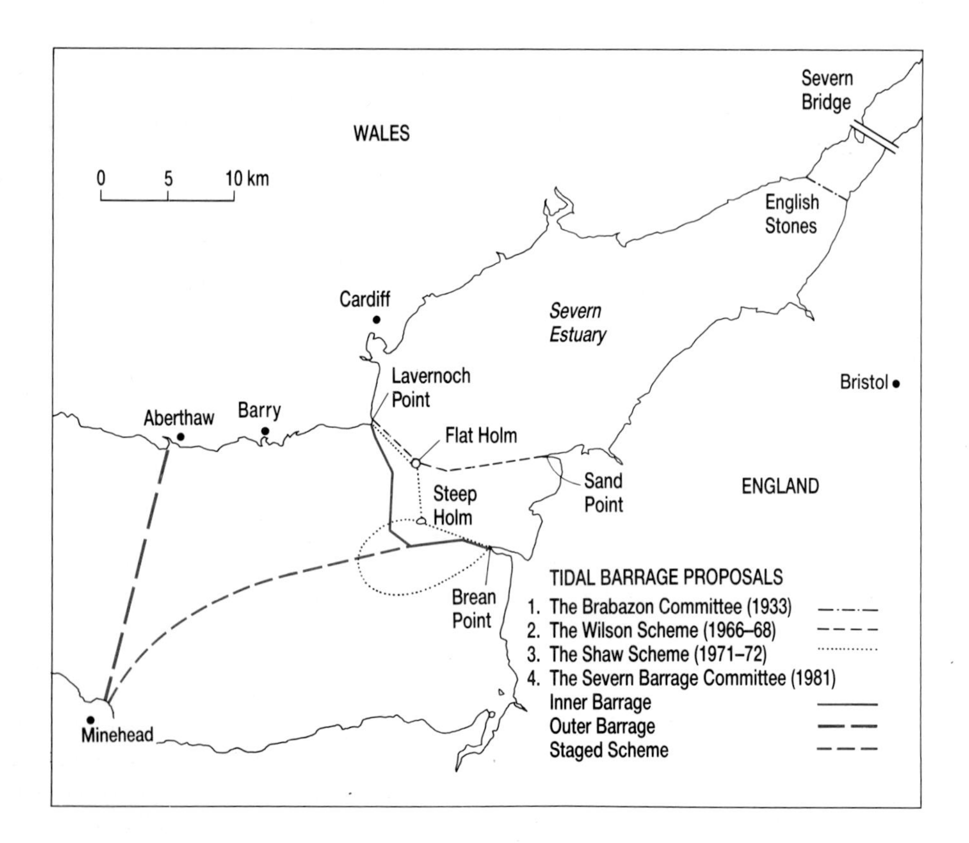

Fig 1.15 Several schemes have been proposed to extract tidal power from the Severn Estuary.

ASSIGNMENT

Fig 1.16 La Rance tidal power station, Brittany.

France's tidal power station

The world's largest and best-known tidal power station is near St Malo in France. The La Rance estuary has been dammed (Fig 1.16), and power generation started in 1966. The output is now 320 MW, provided by 24 sophisticated turbines which have adjustable blades, so that they may operate both when the basin is filling and when it is emptying. You can estimate the efficiency of this station using the information provided in Table 1.2. Fig 1.17 shows how the power output might vary with the rise and fall of the tide.

Table 1.2 Data for La Rance tidal power station

Electrical output	320 MW
Mean tidal range	8.4 m
Area of basin	22 km^2
Density of water	1000 kg m^{-3}

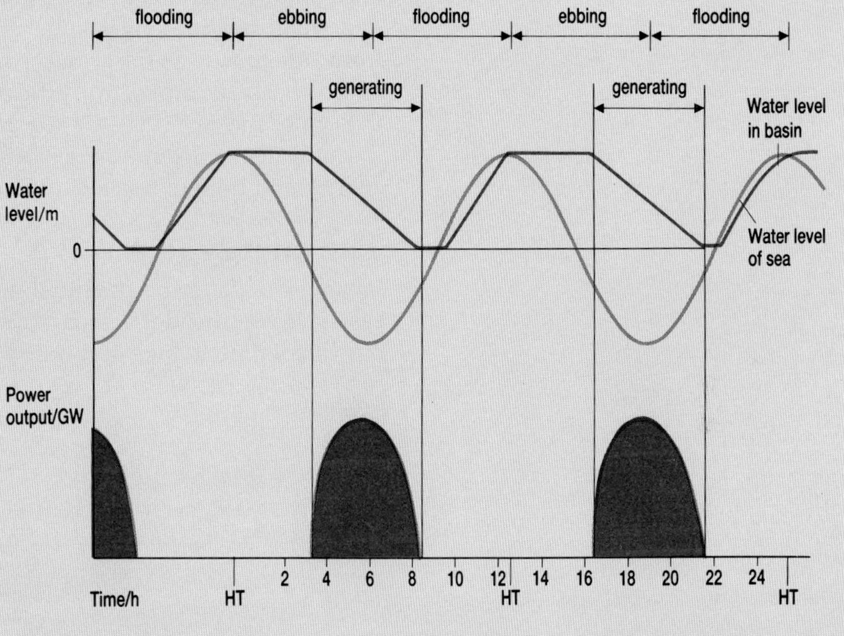

Fig 1.17 Variation of output power for a tidal power station. HT = high tide.

1. Estimate the volume, and hence the mass, of water stored when the basin is full.
2. This water is released when the tide is low. What is the average height of the water above the low water mark? What is its gravitational potential energy?
3. Calculate the efficiency of the entire installation.
4. Why does the power output of the turbines drop rapidly (Fig 1.17) when they are running? How might a more uniform level of power output be achieved?

Wave power

You will be familiar with the idea that waves are a way in which energy may be transmitted from one place to another. The waves on the sea represent a large potential source of energy which might be tapped, and a wide variety of different methods of tapping this energy have been tried.

Let's start our consideration of wave power by developing a physical picture of the energy contained in a water wave. Most of this energy comes from the effect of winds on the sea.

Fig 1.18 shows part of a wave, which we will regard as approximately sinusoidal. You have probably been told that water waves are transverse; that is, the water molecules oscillate up and down as the wave travels

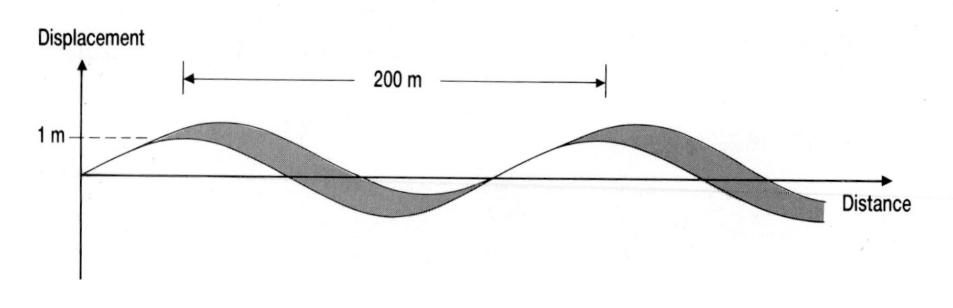

Fig 1.18 A deep sea wave is approximately sinusoidal in form.

horizontally. This is not strictly true, as you will know if you have ever sat in a small boat amongst large waves. The boat moves up and down, and from side to side as the waves pass beneath. These two motions combine to make the boat move in vertical circles, and this is just how the water molecules move.

The water molecules have both kinetic energy and gravitational potential energy, and both of these may be captured by suitable devices.

The energy in a wave

We can calculate the potential energy stored in a wave quite simply. It is easiest if we consider a square wave, as shown in Fig 1.19, and then modify our calculation to take account of the true shape of the wave.

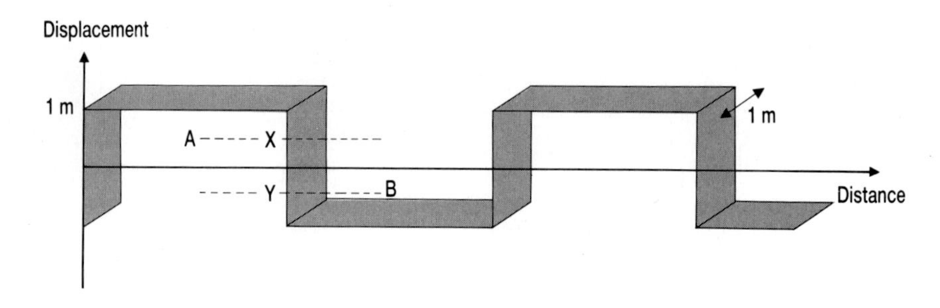

Fig 1.19 A square wave approximation to a water wave.

Our square wave has wavelength 200 m and amplitude 1 m. We are considering a part of the wavefront which is 1 m wide. The section A of the wave has been raised up against gravity; when the energy of the wave has been extracted, section A will fill section B. Its centre of gravity will have moved down from level X to level Y, a vertical distance of 1 m.

To find the change in potential energy of the wave, we need to know its mass, which we can find from its volume and density.

Volume of wave crest = height × width × depth
$= 1 \times 100 \times 1 = 100\ \text{m}^3$
Mass of wave crest = volume × density = $100 \times 1000 = 10^5$ kg
Potential energy of wave crest = mass × g × vertical height
$= 10^5 \times 9.8 \times 1 = 9.8 \times 10^5$ J

If the speed of the wave is 10 m s^{-1}, it follows that one wave arrives every 20 s. The power available from one metre of wave front is thus

power available = energy per second = 4.9×10^4 W.

It turns out that, for a sinusoidal wave, the potential energy is about half that for a square wave. However, waves also have kinetic energy, and this is equal to their potential energy. It follows that our calculation of the available power is a good approximation to the truth. You can see that our calculation shows that, for typical waves on the sea, the available power from one metre of wavefront is about 50 kW.

QUESTIONS

1.1 Show that, for a sinusoidal wave of amplitude A, moving with speed v, the available power P per metre of wavefront is given by $P = \frac{1}{2}A^2Dgv$, where D is the density of water and g is the acceleration due to gravity. (Hint: follow through the logic of our calculation above.)

1.2 Wave power has been suggested as a suitable energy supply for small islands off the coast of the UK. Estimate the length of a wave power station if it is to supply an average of 200 kW to the population of an island. Assume that it operates with 10% efficiency. What problems do you think might be associated with this method of electricity generation? Suggest how these might be overcome.

ASSIGNMENT

Energy from waves

Wave power was the subject of a major UK research and development programme in the late 1970s and early 1980s. Eventually, after a series of reviews, this programme was cut back on the grounds that 'the prospects for large scale offshore wave power were poor compared with other electricity generating renewable energy technologies' – *Taking Power from Water*, Department of Energy, 1989. However, several small scale devices and large scale prototypes were developed, and work continues on some of these. They use a variety of different techniques for extracting energy from waves. Read about some of these devices, and answer the questions which follow.

1. A device such as the Salter Duck is designed to float on the open sea. Estimate the height of the duck. Explain why the surface of the water behind the duck would be smoother than the water in front of it.

Fig 1.20 The Salter duck is perhaps the best known wave energy device. As waves pass, the duck nods, and the relative motion of the two parts may be used to produce electricity, perhaps by forcing hydraulic fluid through a generator. Alternatively, the wave motion might spin gyroscopes within the ducks.

Fig 1.21 The Clam was developed at Coventry Lanchester Polytechnic. It consists of a row of air-filled bags. As a wave passes, the air in the bags is squeezed and forced through a low pressure turbine.

2. Explain why the Clam operates pointing into the waves, rather than lying across them.

3. The energy available from waves is intermittent; it varies as the amplitude and the speed of the waves vary. This is a major problem for wave power. One way that has been suggested to overcome this is to build floating 'factories' alongside the wave power station. Such factories might carry out chemical processes,

working faster when more power is available. Suggest three other ways in which the problem of the variable nature of the output of a wave power station might be overcome.

4. In an oscillating column station, air is driven through a turbine as the waves rise and fall. To make best use of the moving air, it is desirable to rectify this flow, so that it flows the same way through the turbine whether the wave is rising or falling. Write an illustrated account of how you think this might be achieved.

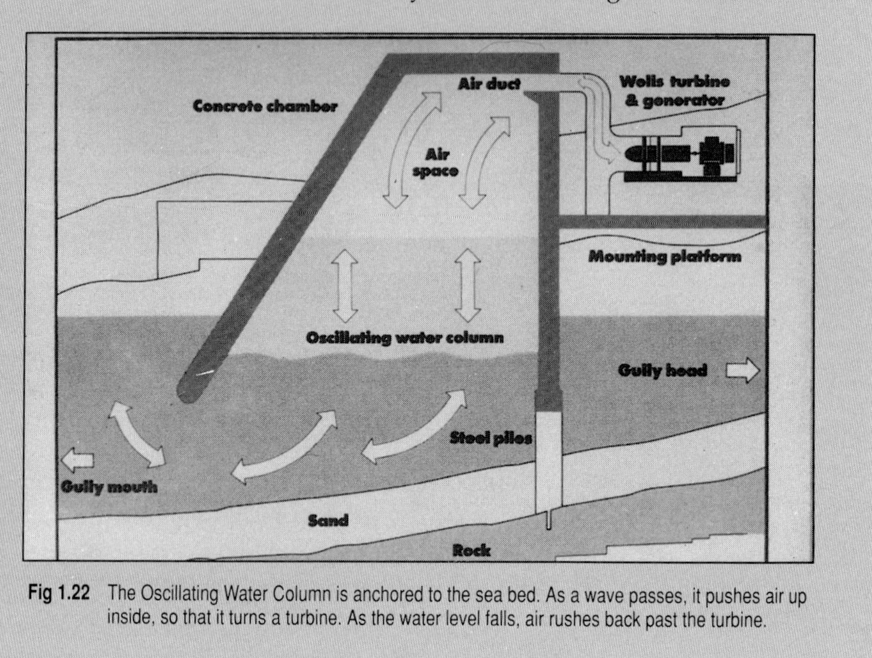

Fig 1.22 The Oscillating Water Column is anchored to the sea bed. As a wave passes, it pushes air up inside, so that it turns a turbine. As the water level falls, air rushes back past the turbine.

1.3 WIND POWER

Much of the energy stored in moving air comes originally from the Sun. The Earth's atmosphere is heated unevenly, and convection currents arise on a global scale. This energy has been harnessed since 2000 BC in the Far and Middle East. It took three thousand years to become widely used in western Europe. All windmills operate on the same principle. Wind blows over a series of blades. It moves faster on one side of a blade, giving rise to a region of low pressure (Bernoulli's Principle). This pressure difference results in a force on the blade, causing it to move.

The horizontal axis wind turbine shown in Fig 1.23(a) obviously bears a close resemblance to a propeller aircraft, and indeed much useful design

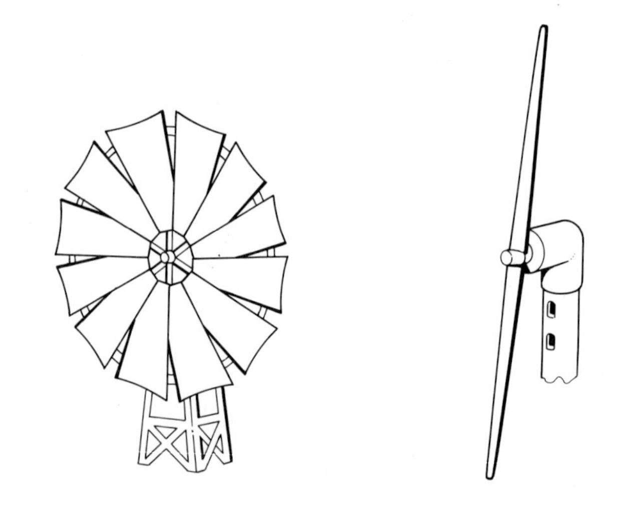

Fig 1.23 Wind turbines: (a) horizontal axis

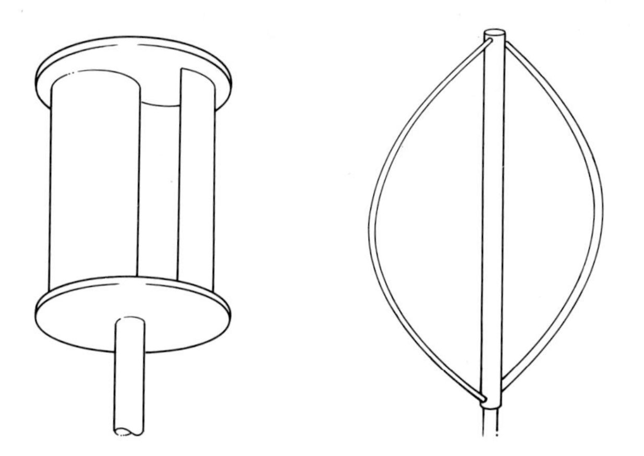

Wind turbines: (b) vertical axis.

information is available from aerodynamics. This has been put to good use in designing early wind-powered electricity generators. However, vertical axis machines (Fig 1.23(b)) are different, and much research is needed to optimise their design. They have the advantage over horizontal axis machines in that they do not need to be turned into the wind as the wind direction changes.

Fig 1.24 The 3 MW aerogenerator at Burgar Hill, Orkney.

Extracting energy from the wind

A 60 m diameter aerogenerator was installed at Burgar Hill, Orkney, in 1987 – Fig 1.24. Its electrical power output is rated at 3 MW. We can estimate the energy available to it when a 20 m s^{-1} wind blows.

The energy of wind is kinetic energy. As the wind blows past the aerogenerator, it is intercepted by the rotating blades which sweep out a circular area. We will calculate the kinetic energy of the air arriving at the generator each second – see Fig 1.25.

Area swept out by blades = $\pi r^2 = \pi \times 30^2 = 2830$ m^2
Volume of air arriving per second = $2830 \times 20 = 56\,600$ m^3 s^{-1}
Mass of air arriving per second = volume × density = $56\,600 \times 1.29$
= 73 000 kg s^{-1}
Kinetic energy per second = $\frac{1}{2}mv^2 = \frac{1}{2} \times 73\,000 \times 20^2 = 14.6$ MW

This is the wind power available to the generator. It is designed to provide 3 MW, its maximum output, when the wind is blowing more slowly than 20 m s^{-1}. When the wind speed increases, the power output will rise rapidly, and the electrical generator might become overloaded. To overcome this, the blades are tilted out of the wind so that the output power returns to 3 MW.

Efficiency

The general expression for the power available to a wind generator is

$$P = \tfrac{1}{2}ADv^3$$

where A is the area swept out by the blades, D is the density of air and v is the wind speed. (Note that the wind speed is cubed in the expression, since the kinetic energy of the wind depends on v^2, and air arrives at the turbine at speed v.)

Of course, not all of this power can be extracted as electrical energy. The fundamental reason for this is that, as the air passes the turbine, it must lose kinetic energy. If it loses all its kinetic energy, it must become stationary. This would prevent any further air reaching the turbine. In practice, the air expands as it slows down beyond the turbine – Fig 1.26. This limits the fraction of the energy which may be extracted to 59% of the total.

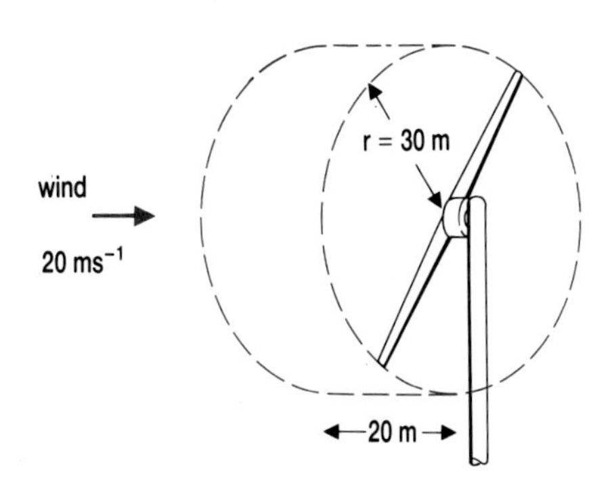

Fig 1.25 Wind arriving at a turbine.

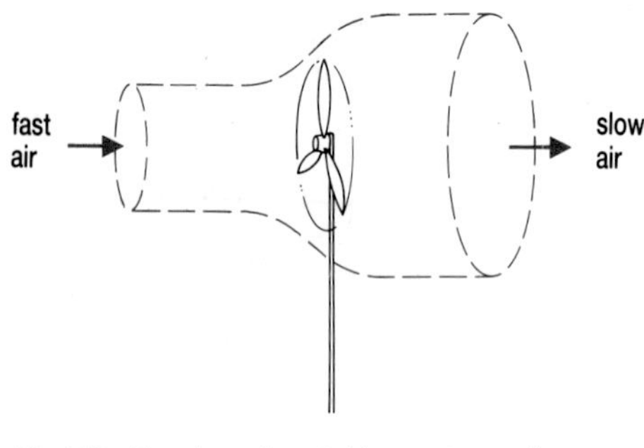

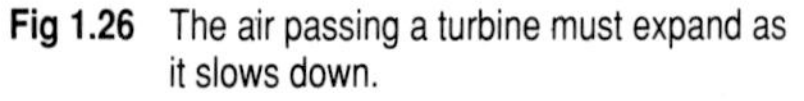

Fig 1.26 The air passing a turbine must expand as it slows down.

Practical turbines with efficient generators can convert about 40% of the wind power to electrical power.

Turning effect

When wind strikes a turbine, it exerts a force on the vertical structure. We can calculate this force and its turning effect by considering the momentum of the air.

We will consider the Burgar Hill station again, with wind arriving at 20 m s^{-1}; the aerogenerator operates with 40% efficiency. Its midpoint is 45 m above the ground.

We have calculated already that the mass of air arriving per second is 73 000 kg s^{-1}, and its kinetic energy is 14.6 MJ. For a 40% efficient turbine, the kinetic energy of the air leaving each second is 14.6 × 0.6 = 8.76 MJ. Now we can find the speed of the air leaving the turbine.

$$\text{speed of air} = \sqrt{2 \times E_k / m} = 15.4 \text{ m s}^{-1}$$

Now the force exerted on the turbine structure is equal to the rate of change of momentum of the air (by Newton's Second Law of Motion).

$$\begin{aligned} \text{force} &= (\text{mass per second}) \times (\text{change in wind speed}) \\ &= 73\,000 \times (20 - 15.4) \\ &= 340\,000 \text{ N} \end{aligned}$$

This force is approximately equal to the weight of an object of mass 34 tonnes, and at a height of 45 m above ground level, it has a large turning effect (moment).

ASSIGNMENT

The Furlmatic wind generator

The Furlmatic – Fig 1.27 – is a wind generator designed for charging batteries. Its output is rated at 24 volts, 250 watts. The graph shown in Fig 1.28 is taken from its technical specification. Use the graph to answer the following questions.

Fig 1.27 The Furlmatic FM1800 wind generator. The diameter of the rotor is 1.8 metres and its maximum output at a wind speed of 28 mph is 24 V, 360 W.

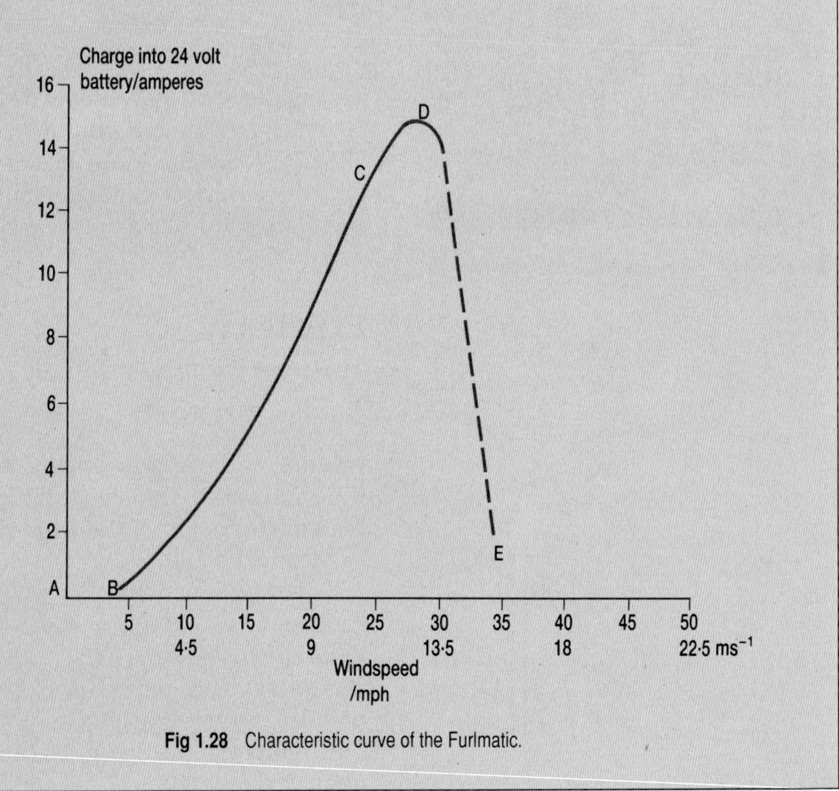

Fig 1.28 Characteristic curve of the Furlmatic.

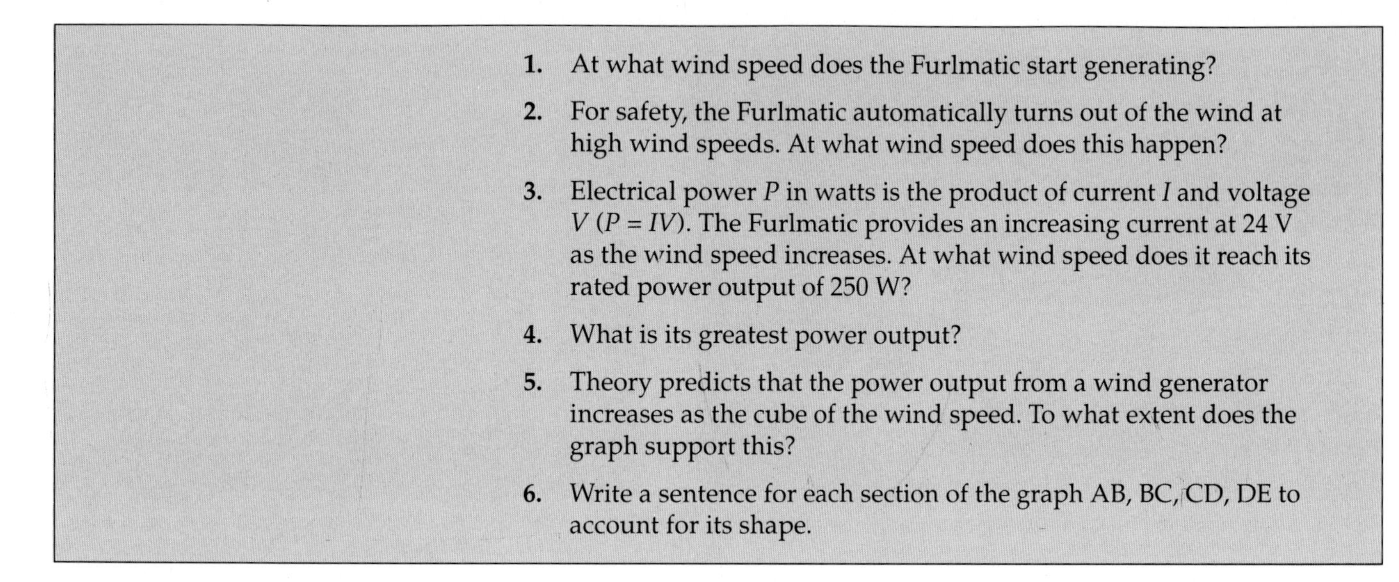

1. At what wind speed does the Furlmatic start generating?
2. For safety, the Furlmatic automatically turns out of the wind at high wind speeds. At what wind speed does this happen?
3. Electrical power P in watts is the product of current I and voltage V ($P = IV$). The Furlmatic provides an increasing current at 24 V as the wind speed increases. At what wind speed does it reach its rated power output of 250 W?
4. What is its greatest power output?
5. Theory predicts that the power output from a wind generator increases as the cube of the wind speed. To what extent does the graph support this?
6. Write a sentence for each section of the graph AB, BC, CD, DE to account for its shape.

1.4 BIOMASS AND FOSSIL FUELS

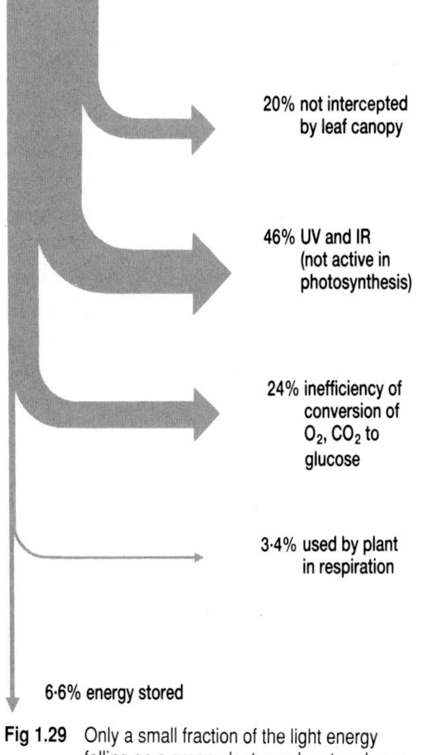

Fig 1.29 Only a small fraction of the light energy falling on a green plant may be stored as a result of photosynthesis.

Table 1.3

Crop	Photosynthetic efficiency (%)
sugarcane	1.2
reed swamp	1.1
grassland	0.8
evergreen forest	0.8
deciduous forest	0.6

In many parts of the world, biological materials provide a major part of our energy supplies. Wood is a vital source of energy for cooking and heating, and collecting wood accounts for up to 40% of the time spent working for people in many African and Asian countries. However, as we will see in this section, there is more to biomass energy than simply collecting wood.

Fossil fuels – coal, oil and gas – make a vital contribution to energy supplies in all economies. They represent a great store of ancient biomass energy. All these fuels originally derive their energy from the process of photosynthesis in living plants.

Photosynthesis

Plants make carbohydrates from carbon dioxide and water. To do this, they need energy which they derive from sunlight. This process is called **photosynthesis**, and it is the principal way in which energy from the Sun is captured by living organisms.

$$\text{carbon dioxide} + \text{water} \xrightarrow{\text{chlorophyll, light}} \text{carbohydrate} + \text{oxygen}$$

Photosynthesis is an inefficient process. Only about 5 or 6% of the energy of the Sun's rays falling on the leaves of a plant might theoretically be stored as recoverable energy in the material of the plant.

Fig 1.29 shows, in the form of a diagram, what happens to the energy falling on a typical plant. This kind of diagram is called a **Sankey diagram**; it shows the forms of energy involved in conversions, together with the relative amounts involved. You will see more Sankey diagrams later in the book, and you will have opportunities to draw some for yourself.

Although photosynthesis is inefficient – compare the theoretical maximum efficiency of photocells, 25% – there are several points to note which make it an attractive proposition. Firstly, the energy is produced in a convenient stored form. This is a great advantage over, say, photocells. Secondly, there are prospects for improving the efficiency of plants. Table 1.3 shows the photosynthetic efficiency achieved by a variety of plant types. They are all well below the theoretical limit. Little effort has been put into improving these figures. Perhaps a concerted programme of plant breeding and genetic engineering could significantly increase the yield, as has been the case with food crops.

Using biomass

Biomass fuels are produced in a variety of forms, and may be converted into other forms for convenient use – see Figs 1.30 – 1.32.

Fig 1.30 A methane generator in a Chinese commune produces both gas and fertiliser from human and animal wastes.

Fig 1.31 Waste materials are increasingly used as an energy source either by being burnt or by providing gas. On this landfill site, however, gas is merely being burnt off in order to avoid a dangerous build-up of methane.

Fig 1.32 The coppicing of trees produces a crop of suitable lengths of wood for charcoal making.

Biogas is largely methane, and is produced by digesting cattle dung and other waste materials in a tank. The gas is piped off for use in cooking and heating.

Vegetable oils and alcohol from sugar have been used for powering vehicles. Gasohol is a mixture of ethanol from fermented plants such as maize and sugar cane, and unleaded petrol. It is an important fuel in parts of North and South America.

Making charcoal is a convenient way of reducing wood to an easily combusted form. The process, called pyrolysis, involves heating wood to a temperature of several hundred degrees Celsius, in the absence of air. Traditionally, it was used in iron-smelting, though it has now been largely replaced by coke. Its most important use is as a cooking fuel.

Some algae consist largely of hydrocarbons. They are especially attractive, as their cells do not have cell walls, which would need to be digested. There are proposals to develop continuous flow methods of growing such algae, where a material like crude oil would be tapped off from tanks of the algae.

Energy in biological materials

The energy stored in fuels is often referred to as chemical energy. It is associated with the chemical bonding of the atoms of which the fuel is composed. In the process of photosynthesis, the atoms are arranged in configurations such that they have high electrostatic potential energy; when the fuel is burnt in the process of combustion, the atomic arrangement is altered to a lower energy state. Heat energy is released.

$$\text{hydrocarbon} + \text{oxygen} \longrightarrow \text{carbon dioxide} + \text{water} + \text{heat}$$

The energy content or **calorific value** of a fuel is a measure of the amount of energy released when the fuel is burnt. Table 1.4 lists the calorific values of a variety of fuels. You will notice that they are all quoted in units of MJ kg^{-1}; that is, the calorific value tells us the amount of energy released when 1 kg of fuel is completely combusted. You may see calorific values quoted in other units, such as MJ m^{-3} for gases. It is usually a simple matter to convert from one unit to another. This will be dealt with in more detail in Chapter 3.

Thermal power stations

In industrial societies, electricity is often the most convenient form of energy supply. Although fossil fuels may be burnt efficiently, technology has developed greatly since the early days of steam engines, and the versatility of electricity means that it has a vast range of uses which cannot be supplied directly by burning coal, oil or gas.

Throughout the world, coal is the major fuel used in electrical power stations. Fig 1.33 shows the principal elements of a coal-fired power station. You should understand the functions of each part. In Chapter 5, you can read about the limits to the efficiency of such a power station, which is generally less than 40%.

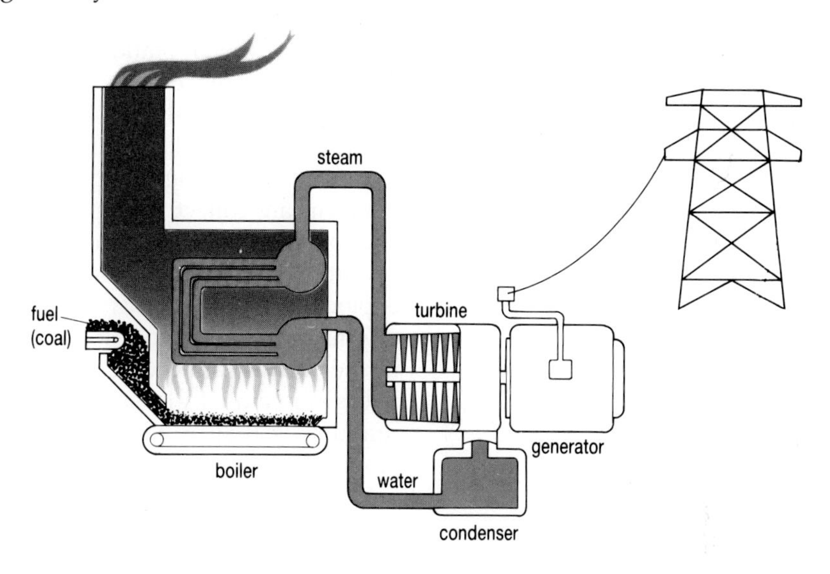

Fig 1.33 The construction of a fossil-fuel burning power station.

ASSIGNMENT

Energy from fuels

The calorific values of fuels shown in Table 1.4 give an indication of the quality of different fuels as sources of heat energy. For example, natural gas provides almost twice as much heat as coal for each kilogram burnt. However, there are other considerations which may be important in the choice of fuels. One is their production of carbon dioxide (CO_2) on combustion. This is a gas which contributes to the greenhouse effect – see Section 3.4. How much CO_2 does each fuel produce?

Table 1.5 lists the approximate carbon content of several fuels. Using the data in these two tables, you can compare the different fuels. A suitable approach is to start by calculating the amount of each fuel needed to provide 1000 MJ of energy. Next, find the carbon content of these amounts. From the relative molecular masses of carbon and CO_2 (12 and 44 respectively), it follows that the amount of CO_2 produced is 44/12 times the mass of carbon in the fuel. Calculate the CO_2 production for each 1000 MJ of energy released, and produce a league table of these fuels.

Table 1.4 Energy content of fuels

Fuel	Calorific value/MJ kg^{-1}
coal (UK average 27)	22–32
oil	42
natural gas	55
wood	20
peat	21
hydrogen	142
carbon	34

Table 1.5 Carbon content of fuels

Fuel	Carbon content/mass %
coal	94
oil	83
natural gas	75
wood	70

(Note: carbon content is as a percentage of the total hydrocarbon content of the fuel.)

1.5 NUCLEAR FUELS

Since the world's first controlled nuclear chain reaction was established in 1942, many nuclear power stations have been built. These rely on the process of nuclear fission, with uranium or plutonium as fuels. Hopes exist that another process, nuclear fusion, may provide a safe and cheap source of energy in the twenty-first century.

Nuclear fission

A nuclear power station is very like a coal-fired power station. The principal difference is in the fuel, and in particular in the way in which energy is stored in that fuel.

Uranium and plutonium are heavy elements. A uranium nucleus is massive, containing 92 protons and over 140 neutrons. (Protons and neutrons are collectively known as nucleons.) These elements, at the heavy end of the Periodic Table, were not present in the early stages of the evolution of the Universe. Rather, they are formed when a dying star explodes as a supernova. The uranium mined on the Earth today had its origins in such an explosion. Part of the energy of the explosion is trapped in the 'bonds' between the particles of the uranium nucleus, and can be released in a nuclear reactor. Thus the difference between coal as a fuel and uranium is that in coal energy is stored in the chemical bonds between atoms, whereas in uranium energy is stored in bonds between nucleons.

How can this energy be released? Uranium is a stable element; the two most common isotopes, U-235 and U-238, have radioactive half-lives of the order of 10^9 years. Uranium has existed since the formation of the Earth, thousands of millions of years ago. To release the energy stored in this fuel, it must be made unstable. The trick is to use neutrons in the process of **nuclear fission**. 'Fission' means 'splitting'.

Fig 1.34 shows how a nucleus of U-235 behaves when it is hit by a slow-moving neutron. Firstly, the neutron is captured. This leaves the nucleus in an unstable condition, and very soon it splits. This is a form of **induced radioactive decay**. In dividing, the nucleus forms two unequal halves or **fission fragments**, and releases two or three more neutrons together with some energy. This energy is mostly kinetic energy of the fission fragments, though some of it appears as kinetic energy of the neutrons and some as gamma rays. As the energy is absorbed by the surrounding uranium, it becomes hot. This heat is extracted and used to boil water in the same way as in a coal-fired power station.

(a)

(b)

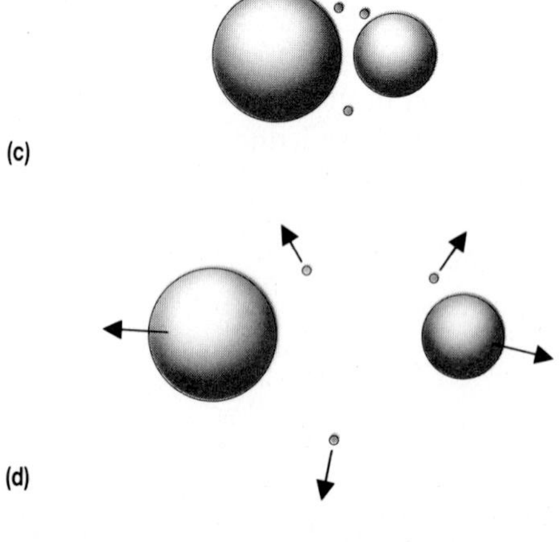

Fig 1.34 A nucleus of uranium-235 undergoes fission after capturing a neutron.

Energy and mass

When a nucleus splits in the process of nuclear fission, it is found that there is a slight loss of mass of the particles involved. We can represent the fission of a uranium nucleus when it absorbs a neutron by an equation:

$$^{235}_{92}U + ^{1}_{0}n \longrightarrow ^{141}_{56}Ba + ^{92}_{36}Kr + 3^{1}_{0}n + Q$$

This equation represents just one of the several different ways in which the nucleus might split. Q represents the energy released.
(Here we are representing a nucleus of element X using the notation

$$^{A}_{Z}X$$

where Z = the number of protons and A = the number of nucleons.)

The masses of the particles shown are given in Table 1.6. They are in **atomic mass units**, for which the symbol is u. 1 u = 1.661×10^{-27} kg. Now we can add up the masses of particles on each side of the equation, and compare the totals:

Table 1.6

Particle	Mass/u
$^{235}_{92}U$	235.043 93
$^{1}_{0}n$	1.008 665
$^{141}_{56}Ba$	140.914 34
$^{92}_{36}Kr$	91.926 25

left hand side	right hand side
235.043 93	140.914 34
1.008 665	91.926 25
	3.025 995
236.052 595	235.866 585

There is a loss of mass in the process of fission

of (236.052 595 – 235.866 585) = 0.186 01 u.

In kg, this is $0.186\,01 \times 1.661 \times 10^{-27} = 3.090 \times 10^{-28}$ kg.

Now, physicists are not very happy to find a process in which mass disappears. Nor do we like to find energy appearing as if from nowhere, apparently breaking the Principle of Conservation of Energy. The simplest way to resolve this problem is to say that the disappearing mass reappears as energy. In other words, mass may be thought of as a form of energy. Then, in the reaction represented by the equation above, the quantity 'energy-including-mass' is conserved.

Albert Einstein devised the equation which allows us to translate between mass in kg and energy in J. This is perhaps the most famous equation in all of science, $E = mc^2$. Here, c is the speed of light in free space, 3.00×10^8 m s^{-1}.

Now we can calculate the energy released in the fission event given by the equation above, since we know how much mass has disappeared.

Mass loss = 3.090×10^{-28} kg

Energy released = $mc^2 = 3.090 \times 10^{-28} \times (3 \times 10^8)^2$

$= 2.78 \times 10^{-11}$ J

This may seem a small amount of energy; however, remember that this is the energy released in the fission of a single uranium nucleus. 1 kg of pure U-235 contains about 2.6×10^{24} atoms, and could thus release approximately 7×10^{13} J of energy if all nuclei underwent fission. Compare this with the calorific value of coal, 29 MJ kg^{-1}, more than a million times smaller.

Thus uranium represents an extremely concentrated store of energy, and is a very attractive proposition as a fuel for power stations. There are, of course, technological problems to be solved before the energy released in the fission of uranium can be made use of in practice, and we will look at these next. But first, to ensure that you have understood how to calculate the energy released in fission, try the question which follows.

QUESTION

1.3 When a U-235 nucleus captures a neutron, it may split in a variety of ways. Equation 1 shows two unequal fission fragments. Here is another equation, in which U-235 splits into two more nearly equal fragments.

$$^{235}_{92}U + ^{1}_{0}n \longrightarrow ^{111}_{44}Ba + ^{122}_{48}Cd + 3^{1}_{0}n + Q$$

Table 1.7 gives the masses of the particles involved. Follow the procedure given above to find the energy Q released in this event.

Table 1.7

Particle	Mass/u
$^{235}_{92}U$	235.043 93
$^{1}_{0}n$	1.008 665
$^{111}_{44}Ba$	110.917 41
$^{122}_{48}Cd$	121.913 52

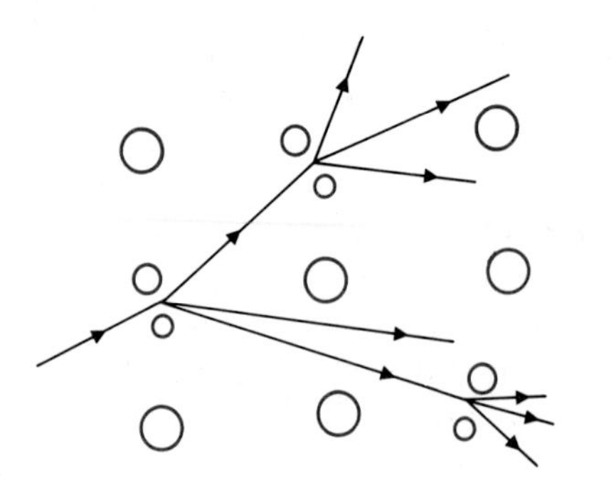

Fig 1.35 A nuclear chain reaction occurs when neutrons released in fission initiate further fission events.

Controlling a chain reaction

To obtain a steady flow of energy from uranium fuel, a **chain reaction** must be set up and controlled. Fig 1.35 shows the principle of a chain reaction. The idea is that at least one of the neutrons released when a uranium nucleus splits goes on to induce fission in another nucleus.

If each neutron induced a further fission, the chain reaction would escalate out of control. In practice, many of the neutrons released are absorbed without causing fission, or they may leave the bulk of the fuel and fail to contribute to the chain reaction – Fig 1.36.

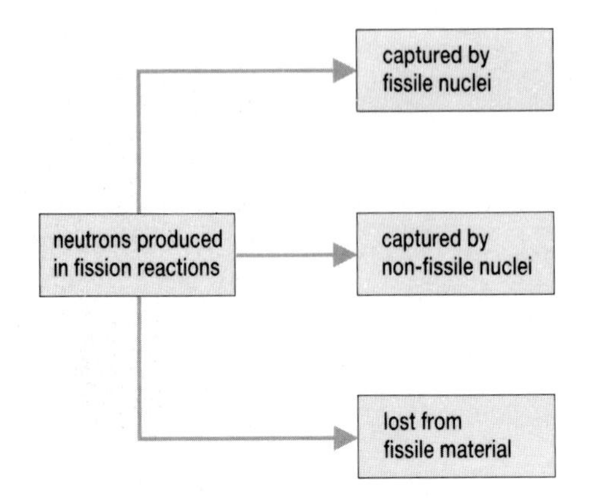

Fig 1.36 The absorption of neutrons released in fission.

Two components in the core of a nuclear reactor (Fig 1.37) are vital for the control of the chain reaction. These are the **moderator** and the **control rods**.

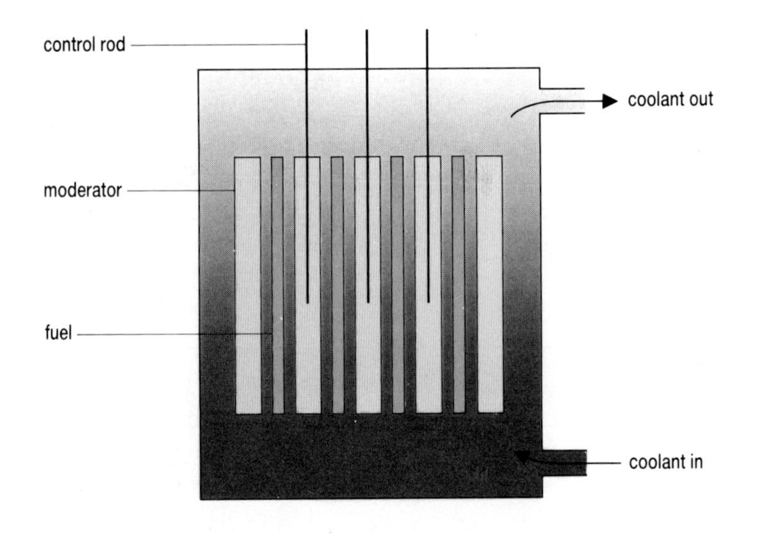

Fig 1.37 The construction of the core of a thermal nuclear reactor.

The moderator is a material, often graphite, included so that it surrounds rods of uranium fuel. Fast neutrons are slowed down as they interact with the moderator; this increases the chance of their being absorbed as they pass close to a uranium nucleus. (A slow neutron is often called a thermal neutron.)

The control rods are made from a material which is very efficient at absorbing neutrons. The flux of neutrons within the reactor core can be increased by withdrawing the rods; if the reaction threatens to start running too fast, the rods are inserted further, and the reaction slows down or stops.

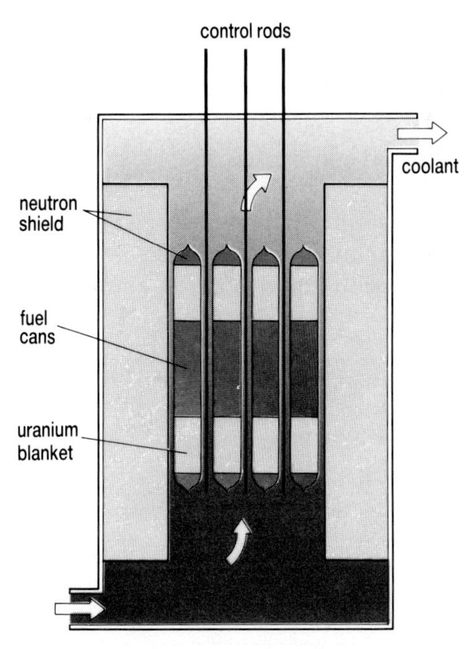

Fig 1.38 The construction of the core of a fast reactor.

Nuclear fuels

Nuclear power reactors use one of two fuels, uranium and plutonium. The useful fissile isotopes are $^{235}_{92}U$ and $^{239}_{94}Pu$. Unfortunately for nuclear technologists, natural uranium contains only about 0.7% of $^{235}_{92}U$; most of the remainder is non-fissile $^{238}_{92}U$. For many reactors, this fuel is too dilute, and it must be enriched to about 3% of the fissile isotope.

Plutonium is a transuranic element, which does not occur naturally. It is made in reactors when $^{238}_{92}U$ captures a neutron, and subsequently emits two beta particles. The resulting $^{239}_{94}Pu$ is used as a fuel for **fast reactors**, and for making some types of nuclear bombs.

Fast reactors are sometimes known as breeder reactors, since they are designed to produce more plutonium while they operate – Fig 1.38. This is done by surrounding the reactor core with a blanket of uranium, which absorbs some of the neutrons which escape from the core. Plutonium is thus created, and can be extracted from the blanket material to act as fuel for a future reactor. Since fast reactors thus make use of the non-fissile $^{238}_{92}U$, they represent a considerably more effective use of uranium, and many nuclear technologists suggest that fast reactors are the best way to make the most of uranium supplies.

Nuclear power stations

Fig 1.39 shows the principal components of a typical nuclear power station. You should notice that most of its construction is the same as for a fossil-fuel power station, Fig 1.33.

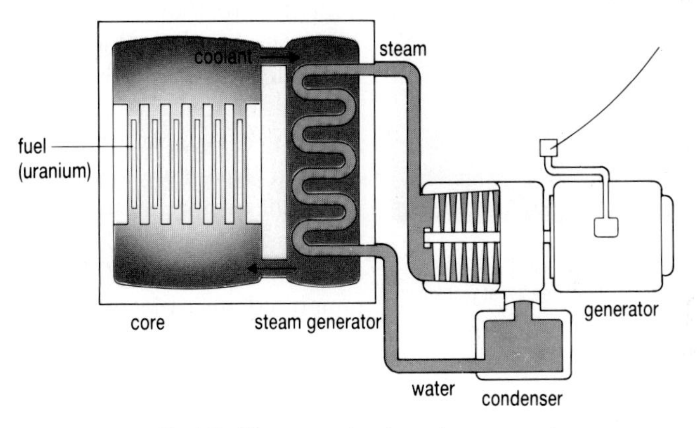

Fig 1.39 The construction of a nuclear power station.

You may have heard of some of the different types of reactors – Magnox, AGR, PWR, CANDU and Fast reactors. The differences lie in the composition of the fuel used, the moderator, and the coolant which carries heat from the core to the boiler. Some use slow (thermal) neutrons, others use fast neutrons. It is not our purpose here to discuss these different types of reactor; you can find out more about them in the literature produced by the electricity generating concerns, and by environmental groups opposed to nuclear power.

Fig 1.40 Fusion of two deuterium nuclei.

Nuclear fusion

When light nuclei join together to form heavier nuclei, they are said to **fuse**. For example, deuterium $^{2}_{1}H$ is an isotope of hydrogen. Two deuterium nuclei (deuterons) can fuse together to form a $^{4}_{2}He$ nucleus, Fig 1.40. In the process, energy is released.

As in the process of nuclear fission, the energy comes from a disappearance of mass. Table 1.8 lists the masses of some light nuclei. We can use these values to determine the energy released in the fusion of two deuterons:

Table 1.8

Particle	Mass/u
${}^1_0$n	1.008 665
${}^2_1$H	2.014 102
${}^3_1$H	3.016 049
${}^4_2$He	4.002 604

$${}^2_1\text{H} + {}^2_1\text{H} \longrightarrow {}^4_2\text{H} + Q$$

The mass of two deuterons is 2 × 2.014 102 = 4.028 204 u. The mass which disappears in the reaction is thus 4.028 204 – 4.002 604 = 0.0256 u, or 4.252 × 10^{-29} kg. Using $E = mc^2$, we can deduce that the energy released, Q, is 3.83 × 10^{-12} J.

Since approximately one in seven thousand hydrogen atoms is the isotope deuterium, and hydrogen is one of the most abundant elements on the Earth, it would seem that deuterium fusion could provide abundant supplies of energy for our future use.

QUESTION

1.4 One reaction suggested for practical reactors is the fusion of deuterium and tritium, ${}^3_1$H.
The equation for this is

$${}^2_1\text{H} + {}^3_1\text{H} \longrightarrow {}^4_2\text{He} + {}^1_0\text{n} + Q$$

Use the data in Table 1.8 to calculate the energy released in this reaction.

Fig 1.41 A view of the inside of the Joint European Torus experimental fusion reactor at Culham near Oxford.

Fusion reactors

Fusion occurs in stars. This is where the elements are built up by the fusing of protons and neutrons to make helium, carbon, and other light elements. The core of a star consists of plasma, a flux of particles, at such a high temperature that fusion can occur. The temperature may be as high as 10^8 K. To emulate this on Earth is the major problem of nuclear fusion technology.

Several experimental fusion reactors have been built, and have achieved some success in producing and containing high temperature plasmas. The Russian Tokamak system and the European JET system (Fig 1.41) use magnetic confinement of the plasma. This relies on the charged nature of the plasma particles, which spiral endlessly around the flux lines of a toroidal (ring-shaped) magnetic field. Fig 1.42 illustrates how the heat generated in a plasma might be extracted in an imaginary fusion power station of the future.

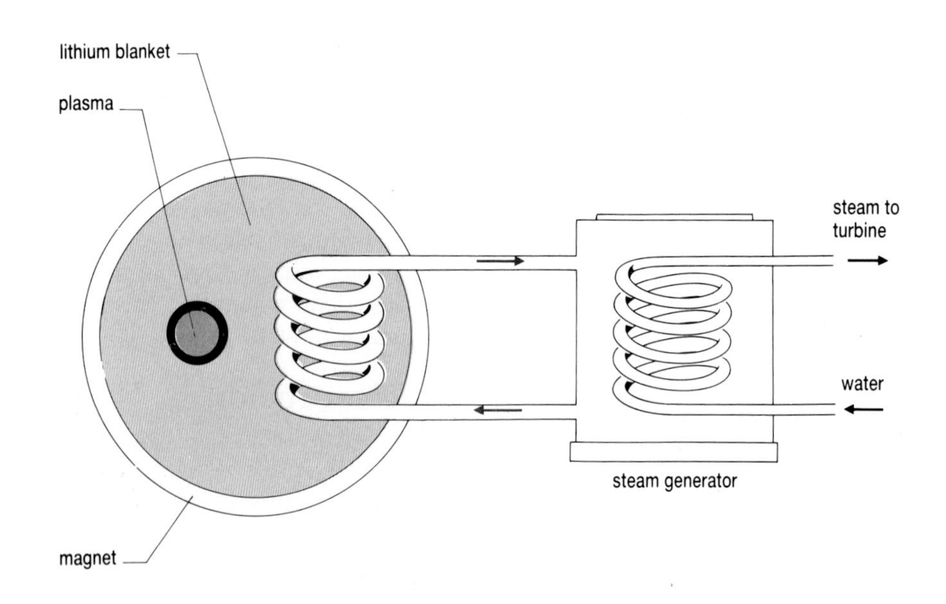

Fig 1.42 The construction of a possible nuclear fusion reactor.

ASSIGNMENT

Water power

Previously, we have compared different fuels by calculating their calorific values in MJ kg^{-1}. The proponents of nuclear fusion power suggest that, since deuterium is a plentiful nucleus in water, fusion could solve all our future energy needs. You can estimate the calorific value of water as a nuclear fuel using the information below.

Fraction of hydrogen nuclei which are ^2_1H = 0.015%
Relative molecular mass of water (H_2O) = 18
Avogadro constant = 6.022×10^{23} particles mol^{-1}
Density of water = 1000 kg m^{-3}
Energy released in fusion of two deuterons = 3.83×10^{-12} J

1. What is the calorific value of water in MJ kg^{-1}?
2. World energy consumption is approximately 10^{21} J per year. How much water would be needed to supply this demand using nuclear fusion reactors?
3. The oceans contain approximately 10^{17} m^3 of water. For how long could this water supply our energy needs at our present rate of consumption?

1.6 GEOTHERMAL AND DEEP OCEAN ENERGY

In some parts of the world, hot water emerges from the Earth as geysers or hot springs. These are usually associated with volcanic activity. More generally, hot water and hot rocks are to be found beneath the Earth's surface. Also, the temperature of the oceans varies with depth. Both of these energy sources have been explored with a view to exploitation.

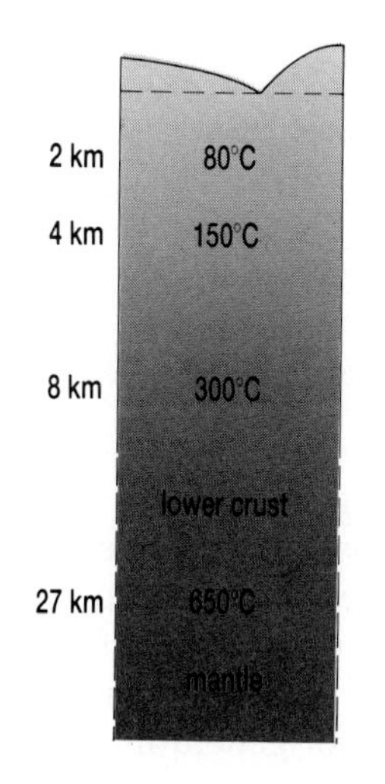

Fig 1.43 The earth's temperature increases with depth.

Geothermal energy

Fig 1.43 shows how the temperature of the Earth's crust increases with depth. The gradient is about 30 °C km^{-1}. Where does this heat come from? The core of the Earth is hot partly because the Earth formed from molten matter, and some residual heat remains. Also, the Earth contains large amounts of radioactive elements such as uranium, thorium and potassium, and the decay of these materials releases heat.

To put geothermal energy in perspective, the rate at which heat reaches the surface of the Earth is about 0.02 W m^{-2}. Compare this with the solar constant (page 2) of approximately 1400 W m^{-2} at the top of the atmosphere, and you will see that we could not rely on heat conducting up through the Earth to keep us warm. However, the Earth has considerable mass, and several technologies exist to extract energy from it in useful quantities.

Hot water is found in porous or fractured rocks at depths of up to 3000 m. This is referred to as an **aquifer**. Aquifers are exploited by pumping the hot water to the surface and using it for heating. Hungary has had homes heated in this way for almost a century; France has a million homes with aquifer heating.

Surveys in the UK have found suitable rocks in several parts of the country – see Fig 1.44 – but exploitation is limited, partly because the water produced is not hot enough to generate electricity. A scheme in Southampton city centre is designed to use water from the Wessex Basin to heat civic buildings. Because the water temperature is only 70 °C, a back-up coal-fired system is needed.

A more practical geothermal energy system relies on extracting energy from **hot dry rocks**. Granite is a rock with a high proportion of radioactive minerals, and their decay enhances the rate of heat flow upwards to the Earth's surface. In places, the rate of heat flow is doubled. Fig 1.44 shows the places in the UK which have large amounts of granite.

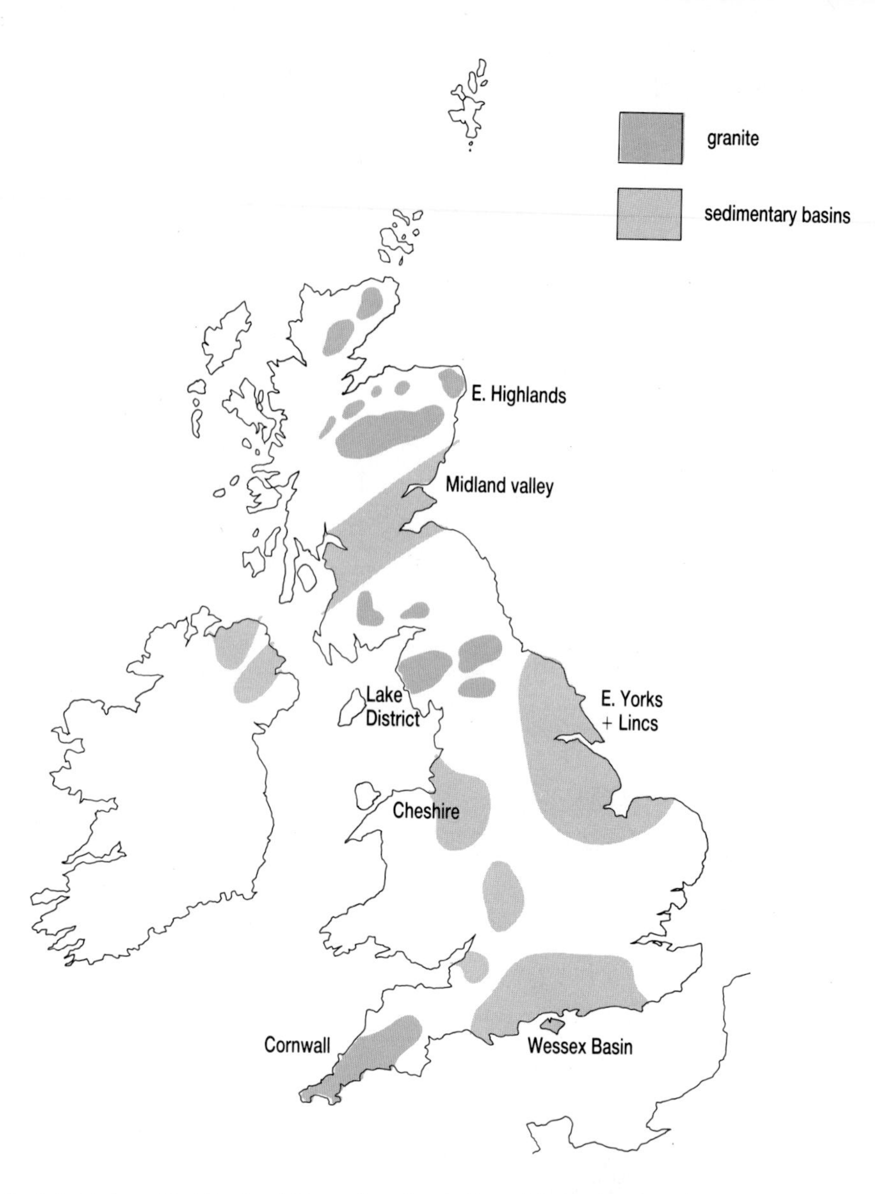

Fig 1.44 Regions of the UK with suitable rock formations for the exploitation of geothermal energy.

the places in the UK which have large amounts of granite.

How can these hot but dry and impermeable rocks be exploited? Fig 1.45 shows the principle of a geothermal (HDR) power station. Two shafts are drilled down into the rock, and explosive charges detonated to fracture the rock. Water is then pumped down one shaft, and hot water is forced up through the other. Water above 150 °C is suitable for generating electricity; temperatures of around 300 °C are possible.

Such a 'doublet' might produce energy at a rate of 50 MW; however, since the temperature is low, only 5 MW of electrical power could result, and the residual hot water would be used for heating.

Deep ocean energy

Wherever a temperature difference exists, there is potential for extracting energy. As we will discuss in Chapter 4, the bigger the difference, the more efficient it is to use the energy to do work.

There are temperature differences in the sea; in particular, the surface layers of tropical oceans are rarely cooler than 25 °C, whilst lower down the temperature is close to freezing. In Ocean Thermal Energy Conversion (OTEC), this temperature difference is exploited. Fig 1.46 shows the principle.

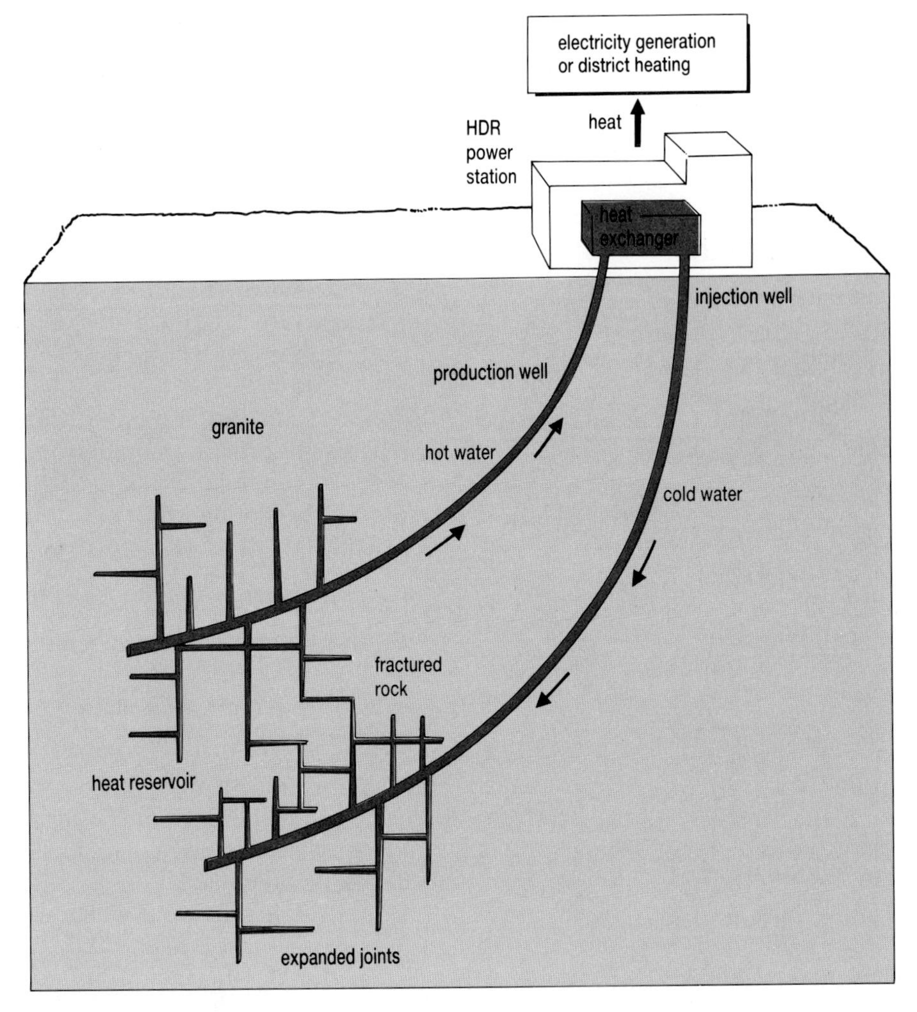

Fig 1.45 The construction of a geothermal power station.

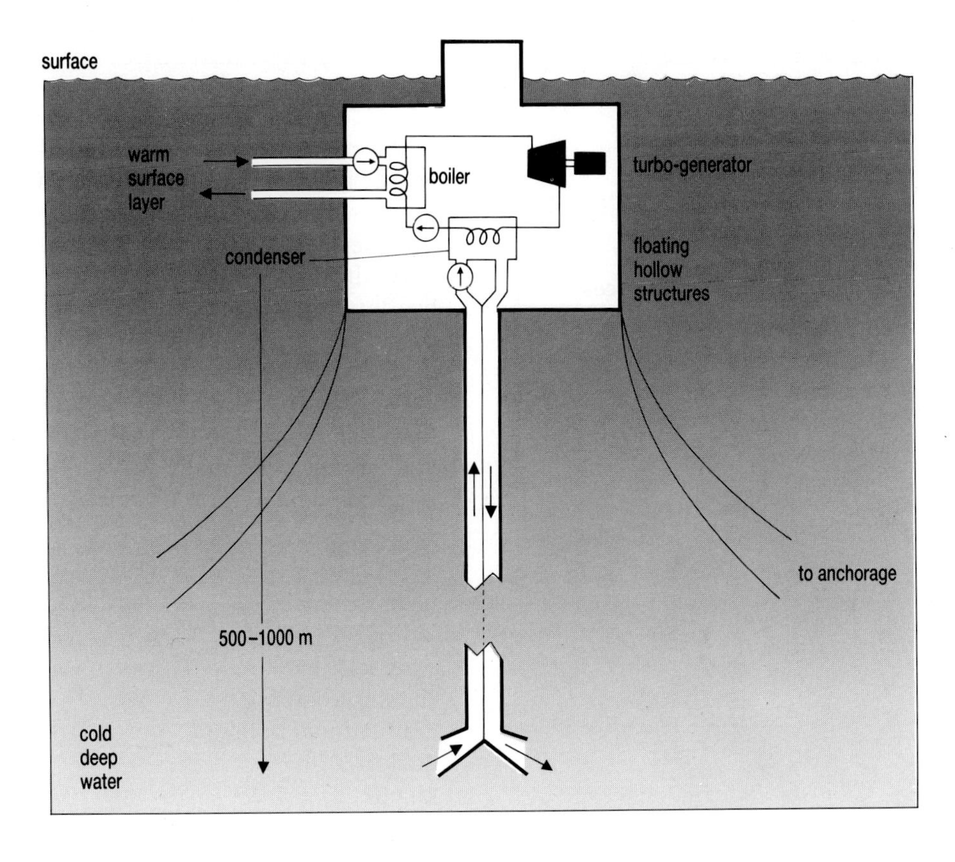

Fig 1.46 The construction of an OTEC power station.

Because the temperature difference between the boiler and the condenser is small compared to a coal-fired power station, the efficiency is low, of the order of 4%. However, experimental systems have proved themselves capable of generating more electricity than is required to operate the pumps for circulating the water. Fig 1.47 shows one such experimental OTEC power station.

ASSIGNMENT

Hot rocks energy

How feasible are hot dry rock and ocean thermal energy conversion systems? What volumes of water must be pumped through to extract reasonable quantities of energy?

Table 1.9

Specific heat capacities	
water	4200 J kg^{-1} K^{-1}
granite	820 J kg^{-1} K^{-1}
Densities	
water	1000 kg m^{-3}
granite	2640 kg m^{-3}

Consider a HDR station producing heat energy at a rate of 50 MW. (This might provide enough heat for a town of 50 000 people.) Cold water is pumped down at 20 °C, and emerges under pressure at 300 °C. The station extracts heat from a volume of 1 km^3 of granite. Use the data in Table 1.9 to answer the questions which follow.

1. How much water must be pumped underground each second?
2. By how much does the temperature of the rock fall in one year?

Now consider an OTEC station producing heat at the same rate. It cools warm water from 25 °C to 23 °C.

3. How much water must be pumped through the boiler each second? If the system has an efficiency of 3%, how much water must be pumped through per second to generate 50 MW of electrical power?
4. Calculate the rate at which water must flow through a 50 MW hydroelectric power station if it has a head of 300 m and operates with 80% efficiency. Compare this answer with your answers to questions 1 and 3.

1.7 OVERVIEW : ORIGINS AND TRANSFORMATIONS

In this chapter, you have studied the physical principles behind a variety of energy sources. The assignment will ask you to take a look at these sources to assess their general origins, and the energy transfers involved. Table 1.10 lists the energy sources we have considered.

We draw on a variety of energy sources. There is increasing concern that, by drawing on ancient reserves, we may be making profligate use of precious energy supplies. To achieve a constant supply of energy, it may be desirable to use renewable resources wherever possible.

We cannot create or destroy energy; we can only transform it from one form to another. Table 1.11 lists a variety of forms of energy. In this chapter, you have studied many ways in which energy is transformed to produce a more convenient form for our use.

Table 1.10 Energy sources
solar heating
solar electricity
wave power
hydroelectricity
tidal power
wind power
biomass
biogas
coal, oil, gas
nuclear fission
nuclear fusion
geothermal
ocean thermal

Table 1.11 Forms of energy
heat (thermal)
light
electrical
chemical potential
gravitational potential
nuclear
kinetic
sound

ASSIGNMENT

Energy origins and transfers

Consider the energy sources listed in Table 1.10.

1. Which forms are derived from recent solar radiation? Which derive from ancient solar radiation? Which have their origins in the formation of the Earth, $4\frac{1}{2}$ billion years ago?
2. Which forms of energy are renewable, and which are non-renewable? Try to justify your answer in each case. Present your answers to these questions in a suitable visual form.
3. For each of the energy sources listed in Table 1.10, what are the initial and final forms of energy? What device is used to make the transformation? Remember that there may be several forms at the beginning or at the end. There may be intermediate forms also. Present your answers in a suitable visual form.

SUMMARY

Most of the energy available to us derives originally from radiation from the Sun. In order to capture this energy and convert it to more useful forms, a variety of conversion devices have been developed. Knowing the physical principles of these devices allows us to understand how they work, and how they can be made best use of.

Chapter 2

ENERGY CONSUMPTION

We have seen that there are a variety of different energy supplies available to us. Our choice depends on which we find most convenient, and which makes best use of resources. We pay for convenience, since convenient energy supplies such as electricity are generally the least efficient in their production from natural resources.

We live in an age when it is increasingly necessary to make efficient use of energy supplies. In this chapter, we will take a close look at electricity distribution and control, and at the conservation of energy in buildings.

LEARNING OBJECTIVES

After studying this chapter you should be able to:

1. explain the principles of a grid electricity supply system;
2. explain the principles of transformers as used in the electricity supply industry, and calculate their efficiency;
3. describe ways in which the electricity industry copes with variations in demand for power;
4. use U-values in calculating heat losses by conduction from buildings;
5. outline the principles of district heating and combined heat and power systems.

2.1 ELECTRICITY DISTRIBUTION

In Chapter 1, we saw that electrical energy may be generated from a variety of sources. These technologies have been developed because of the great convenience and versatility of electricity. In industrialised countries, it is available at the flick of a switch; it can be used for a great variety of tasks. It is hard to imagine how else we might power a food-processor, a computer or a tropical fish-tank!

Although it is possible to generate electricity locally for particular tasks, most of the electrical energy supplied in a country comes from a national grid. The United Kingdom has a grid – Fig 2.1 – which allows the transfer of energy from one place to another as demand varies. Other European countries are similarly linked in major groupings, shown in Fig 2.2. It is interesting to realise that, within such international grid systems, all power stations must produce alternating current of the same frequency and phase. Thus all generators must rotate in step with one another.

Grid technology

Electrical energy is transmitted through a **grid system** using alternating current. This has a number of advantages over direct current; in particular, it allows the use of transformers for changing voltages to meet different requirements. Fig 2.3 shows schematically some typical transformations of voltage between power station and end-user.

High voltages are used for transmission, since this reduces the current which must flow. From the equation $P = VI$, it follows that to transfer

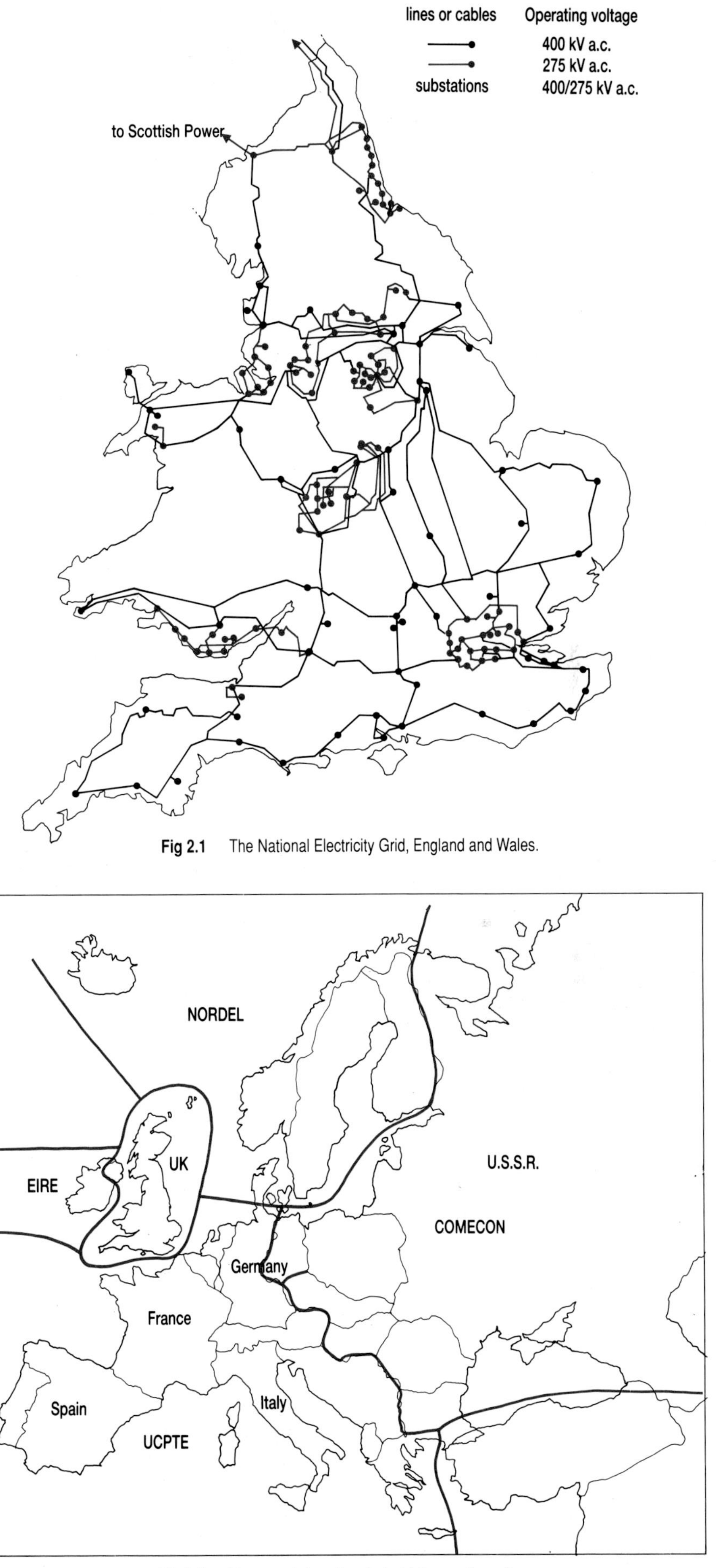

Fig 2.1 The National Electricity Grid, England and Wales.

Fig 2.2 European international electricity supply pools.

power P, the greater the voltage, the smaller the current. The power lost in the transmission cables is given by $P = I^2R$, where R is the total resistance of the cables; from this it follows that the smaller the current, the less is the power loss.

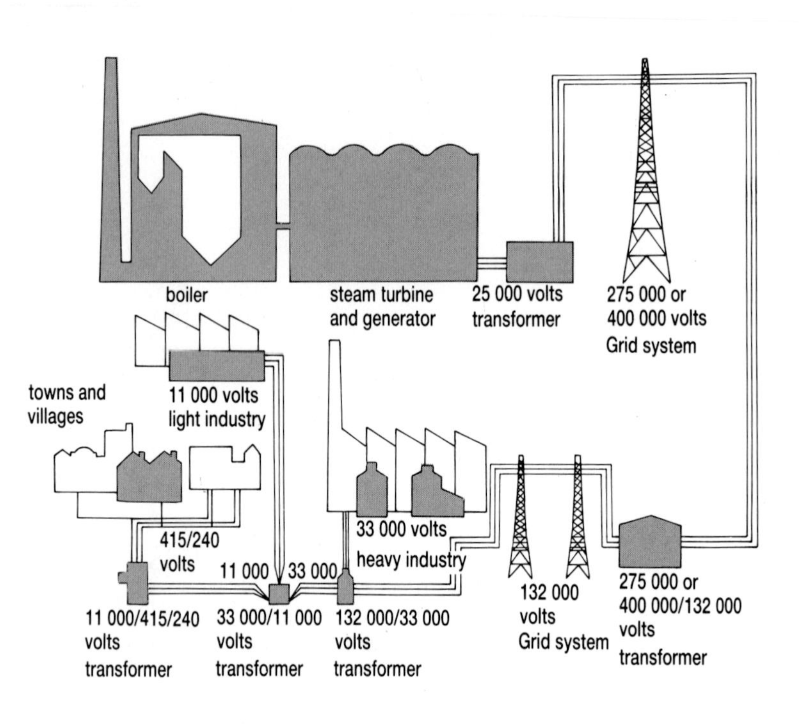

Fig 2.3 Typical voltages used in electricity supply.

Fig 2.4 The construction of a transformer.

The principle of a **transformer** is as follows (see Fig 2.4): two coils, the primary and the secondary, are wound around a soft iron core. An alternating voltage makes a current flow in the primary coil. This produces alternating magnetic flux in the core, which is transmitted to the secondary coil. This changing flux induces an e.m.f. across the secondary coil, which can then drive a current around the secondary circuit.

A step-up transformer increases the voltage; it has more turns on the secondary than on the primary. A step-down transformer is the reverse. For a perfectly efficient transformer with no power losses, the current I_p and p.d. V_p are related to the current I_s and e.m.f. E_s by

$$I_pV_p = I_sE_s$$

and since the voltage is stepped up in the turns ratio of the transformer, it follows that

$$E_s/V_p = I_p/I_s = N_s/N_p$$

where N_s and N_p are the number of turns of wire in the secondary and primary coil, respectively.

Transformers themselves cannot be 100% efficient. There are losses in the primary and secondary windings, and due to eddy currents in the core linking the windings. There are vibrations, and sound is produced. Great improvements in transformer design have been made during the last 100 years, and losses in a well-designed transformer can be less than 1%. Reductions in transformer losses result in great savings, because fewer power stations need to be built.

You can find out more about the value of transformers in the investigations which follow.

INVESTIGATION

A model grid system

Fig 2.5 shows a model system in which power is transmitted along cables to a lamp. You should try this first without the step-up and step-down transformers; the lamp will not light. Use a voltmeter to investigate where the losses occur. What current is flowing through the lamp?

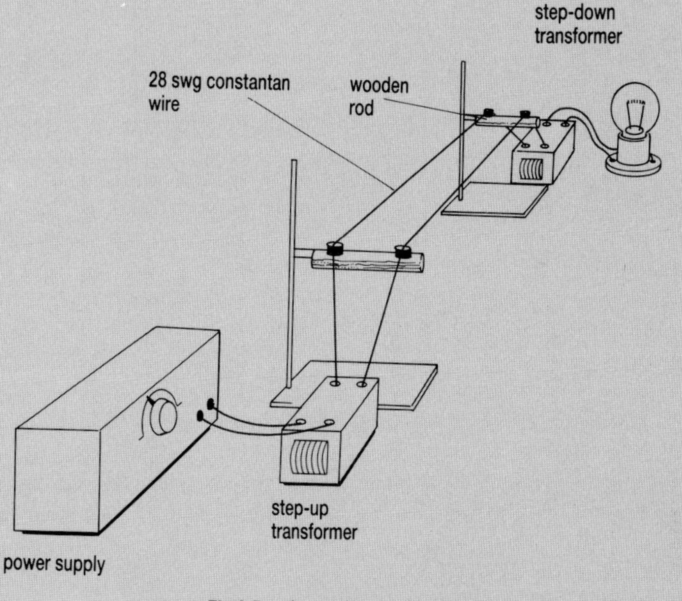

Fig 2.5 A model power line system.

Now repeat the experiment, this time with the transformers.

NOTE: You must not use voltages greater than 50 V. Use the transformer equation to ensure that you do not exceed this value. The lamp should light. Measure and record voltages between, and currents at, appropriate points in the circuit.

WARNING: Switch off when connecting meters in this circuit. Do not touch the power lines when the power supply is switched on. Stand behind a safety screen before switching on.

Calculate the efficiency (power out / power in) for each transformer, and for the transmission lines. Where are the greatest losses? What is the overall efficiency of the system?

INVESTIGATION

The efficiency of a transformer

You can investigate the efficiency of a transformer using a circuit such as is shown in Fig 2.6. You can change the frequency and voltage of the supply using the controls of the signal generator. You can change the load on the system by changing the number of lamps connected (in series or parallel) to the secondary coil.

To determine the transformer's efficiency, you will need to measure the power supplied to the primary windings, and the power leaving the secondary. To do this, you can use joule or watt meters, or use ammeters and voltmeters to find currents and voltages. Then use $P = VI$ to calculate the power.
How does the efficiency of the transformer depend on supply

voltage? How does it depend on the supply frequency? Is the transformer more or less efficient when the demand for power is greater?

What forms of energy loss can you detect in the transformer?

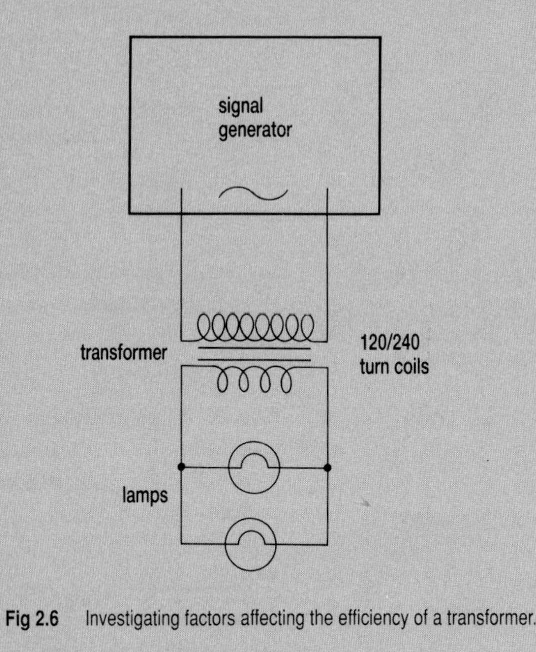

Fig 2.6 Investigating factors affecting the efficiency of a transformer.

ASSIGNMENT

Transformer efficiencies

The electricity industry uses a range of transformers of different sizes. The largest are associated with power stations and the national Super Grid; smaller ones are needed for transforming smaller amounts of power for local uses. How does transformer efficiency depend on the size of a transformer?

Table 2.1 shows data for some typical electricity board transformers of different sizes. The transformers with the biggest power ratings are also those which are physically the biggest.

1. Calculate the efficiencies of these transformers using the equation

$$\text{efficiency} = \frac{\text{input power} - \text{loss}}{\text{input power}} \times 100\%$$

2. What can you say about how the efficiency of these commercial transformers depends on their sizes?

Table 2.1 Data for commercial transformers

Power rating MW	Voltage rating HV/LV kV	Load loss kW
0.1	11/0.433	2.0
0.5	11/0.433	6.9
1.0	11/3.3	12.0
30	132/11	230
60	132/33	390
90	132/33	580
240	400/132	980
500	400/275	730
600	432/22	1200
750	400/275	840

2.2 CHANGING DEMAND

The electricity supply industry has to cope with a demand for power which can show dramatic fluctuations, both in the short term and in the long term. In the assignment which follows, you are asked to think about the reasons for these changes. Then we will go on to look at how the industry operates in practice to cope with this variable demand.

ASSIGNMENT

Variations in demand

Fig 2.7 shows how the total demand for power from the Central Electricity Generating Board varied in 1987. It includes typical summer and winter days, and the days of maximum and minimum demand during the year.

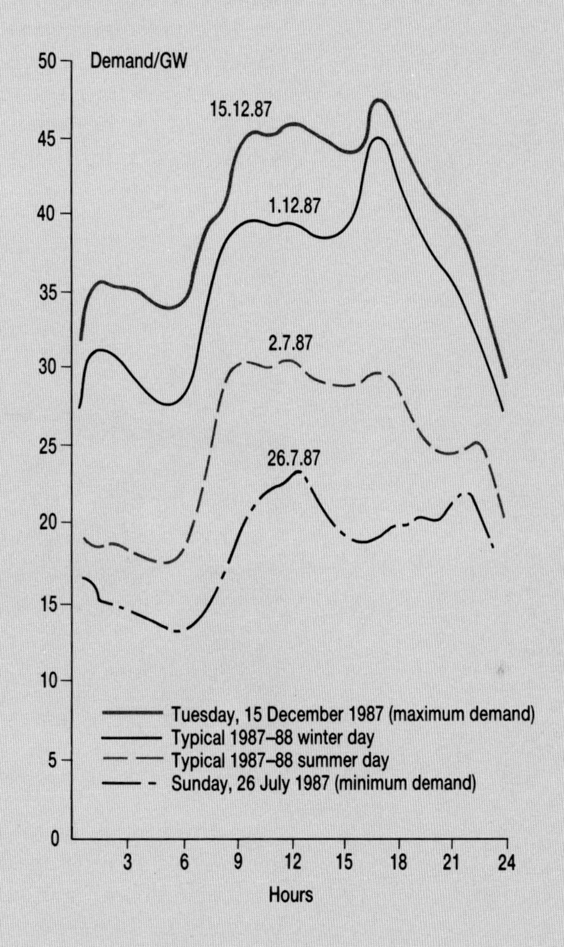

Fig 2.7 Variations in electricity demand in the UK, 1987.

Study the graph, and then answer the questions which follow.

1. By how much does demand vary
   (a) during the course of a typical winter day,
   (b) during the course of a year?
2. A large power station produces about 1000 MW (1 GW) of electrical power. By how much does the number of power stations required to meet demand vary between periods of high and low demand?
3. What reasons can you suggest for the rises and falls in demand in the course of a typical winter day? Why is the pattern different for a typical summer day?

Coping with variations

Changes in demand provide a severe test for the electricity supply industry. An increased load on the system results in a sudden drop in the supply frequency; the voltage may also drop slightly. Many electric motors and other devices derive their speed of operation from the mains frequency. If the frequency drops because of increased demand, their speed drops. It is as if the whole country slows down!

To overcome this problem, changes in demand must be anticipated. Additional power stations must be operating in readiness for a sudden increase – this is known as 'spinning reserve'. A controller for the whole network sits in a control room (Fig 2.8) surrounded by television sets and with copies of Radio and TV Times to hand. This is not because he has nothing better to do with his time; rather, it is because demand is related to how the population are spending their time. If a large number of people switch off their television sets after a popular programme, demand increases for two reasons. Firstly, kettles are turned on, and secondly, people go to the toilet. This puts great demand on the water industry, which must pump large amounts of water and sewage in a very short period of time.

Fig 2.8 The National Grid Control Room.

In deciding how to make best use of available generating capacity, several factors must be taken into account. These include how expensive a particular power station is to run, and how quickly it can be started up. Typically in the UK, nuclear power stations and some large, modern coal-fired stations provide the 'base load', as they have relatively low operating costs and take several hours to bring up to full production. Smaller, older coal-fired stations are less efficient, and are reserved for use in the winter, at times of peak demand.

Another way to cope with the demands placed on the system, is to buy power from the continent. It is supplied via a d.c. link across the English Channel. This has proved to be a cheaper way of coping with demand than building additional power stations.

Sudden increases in demand are coped with in a different way. **Pumped storage systems** have been developed which can provide large amounts of power within a matter of seconds. Fig 2.9 shows the principle of one such system, at Loch Awe in the West Highlands of Scotland. At times of high demand, water floods down from the upper reservoir, to produce up to 400 MW of electrical power. When demand is lower, perhaps at night, the generators work in reverse to pump water up from Loch Awe to the reservoir.

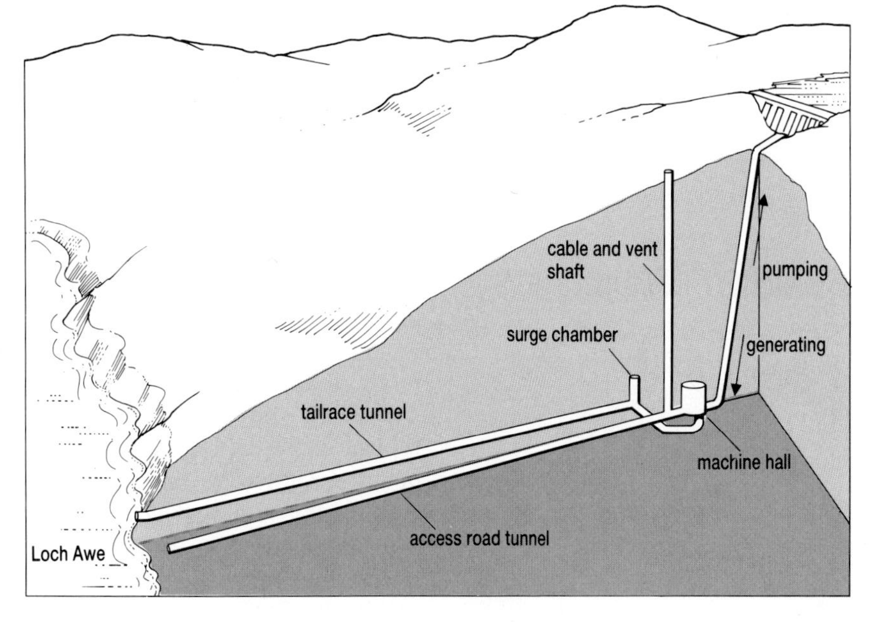

Fig 2.9 The Loch Awe pumped storage scheme.

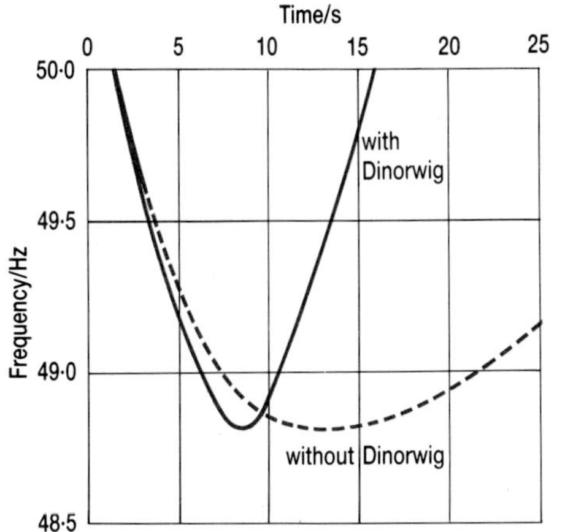

Fig 2.10 Frequency changes due to sudden changes in electricity demand are compensated for rapidly by the use of a pumped storage system.

Such schemes have been built in mountainous regions of Scotland, Wales and Northern Ireland. Fig 2.10 shows how the Dinorwig pumped storage scheme can speed up the return to the correct system frequency of 50 Hz when demand changes suddenly. The generators can be run up to speed using compressed air when an increase in demand is anticipated, so that only a few seconds are needed to restore the system frequency when the water is released from the reservoir.

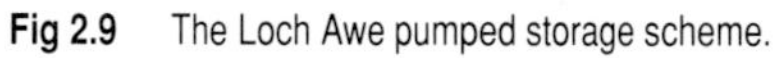

INVESTIGATION

Changing electrical demand

You can investigate the changes which result from a change in load on an electrical system using the apparatus shown in Fig 2.11. An electric motor is used to turn a second motor, which acts as a generator. This supplies electrical power to a lamp. (You will need to take care that you match the lamp to the generator, so that the voltage supplied is enough to light the lamp.)

With the system running, switch in a second lamp in parallel with the first. This increases the demand in the system. What happens to (a) the frequency of the generator, and (b) the voltage and current it supplies? Devise methods of measuring these quantities, and investigate how they depend on the number of lamps in the circuit.

Investigate how you can return the system to its original frequency and voltage by changing the power supply.

In your account of this investigation, explain how this model system corresponds to a full-scale electricity supply system. What does it tell you about the effects of changing demand on a real system?

Fig 2.11 A model system for investigating changes in electricity demand.

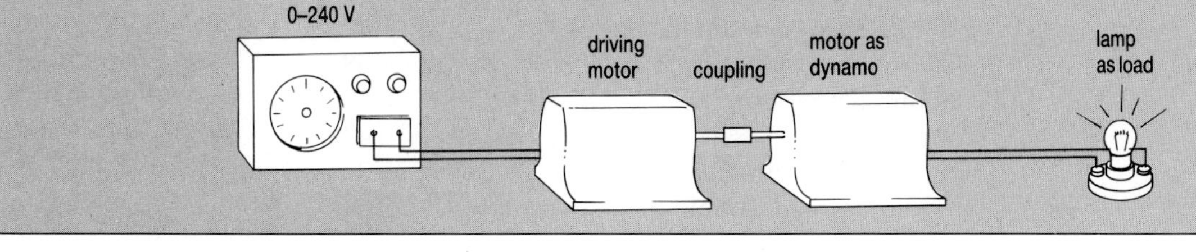

QUESTIONS

2.1 Many homes use 'Economy 7' electricity. For seven hours at night, electricity can be bought at a greatly reduced price.
(a) Why might this benefit the electricity generating industry?
(b) Why is this useful for the consumer?
(c) What evidence is there in Fig 2.7 that significant use is made of Economy 7?
(d) Why is there no corresponding cheap rate for gas?

2.3 ENERGY IN BUILDINGS

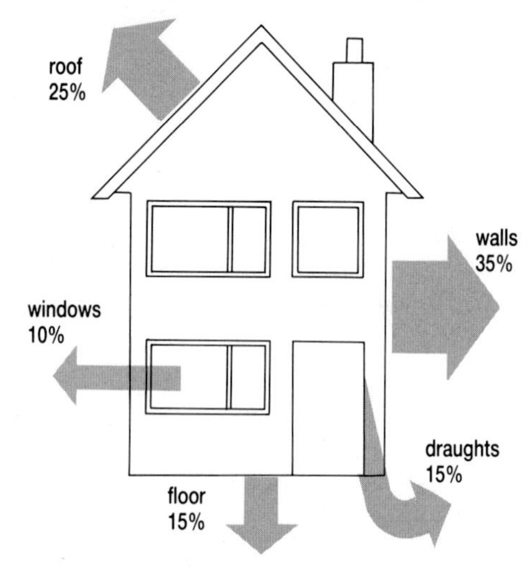

Fig 2.12 Heat losses from a house.

A great proportion of the energy consumed in industrialised societies is used for **space heating** – making habitable the places where we live and work by warming them. It is easier to relax, or to work productively, in a room which is comfortably warm.

Fig 2.12 shows the proportions of heat lost from different parts of a typical house. In Chapter 1, we discussed ways in which buildings may be better designed to make use of available energy from sunlight. In this section, we will look at the ways in which energy in buildings can be conserved by better design, and also at some other ways of providing heat which can make more efficient use of fuel supplies for space heating in towns and cities.

Thermal conduction

Much of the heat lost from a building is conducted away through the floor, walls, windows and roof. Installing suitable insulation materials and double-glazing can reduce this dramatically.

The rate at which heat flows through a material depends on the material, its area A and thickness l, and the temperature difference across it ΔT:

$$dH/dt = kA\Delta T/l$$

where k is the thermal conductivity of the material. In practice, builders and architects who are dealing with materials of known thickness consider U-values rather than thermal conductivities:

$$\text{rate of heat flow} = \text{U-value} \times \text{area} \times \text{temperature difference}$$

$$dH/dt = UA\Delta T$$

The **U-value** for a particular construction material tells us the rate at which heat is conducted away through 1 m^2 of the material for each 1 degree difference in temperature between the inside and the outside of the building. It is much simpler to use U-values in calculations rather than values of k, since a wall or roofing material may consist of several materials together, perhaps brick, plaster and air. Some typical values are given in Table 2.2. From this data, you should be able to see the advantages of installing cavity-wall insulation, loft insulation and double-glazing.

Table 2.2 U-values of typical construction materials

Material	U-value/W m^{-2} K^{-1}
brick cavity wall, no foam insulation	0.77
cavity wall with foam insulation	0.42
slate roof over plasterboard	1.58
100 mm glass fibre	0.39
concrete floor	0.42
single-glazed window	5.3
wooden framed double-glazed window	3.0

Degree days

In calculating energy requirements for heating homes and other buildings, it is necessary to take account of temperature variations from place to place. To do this, the idea of the **degree day** has been developed. It is considered that a comfortable working temperature indoors is 18.5 °C. If the temperature outside falls below 15.5 °C, activity indoors will not be enough to maintain this working temperature, and heating will be needed. The further the temperature drops, the more heating will be needed. If the average outside temperature is, say, 13.5 °C, then two degree days of heating are needed.

Fig 2.13 shows the number of degree days per year in different regions of the UK. These were obtained by averaging measurements over the 20 years to 1987.

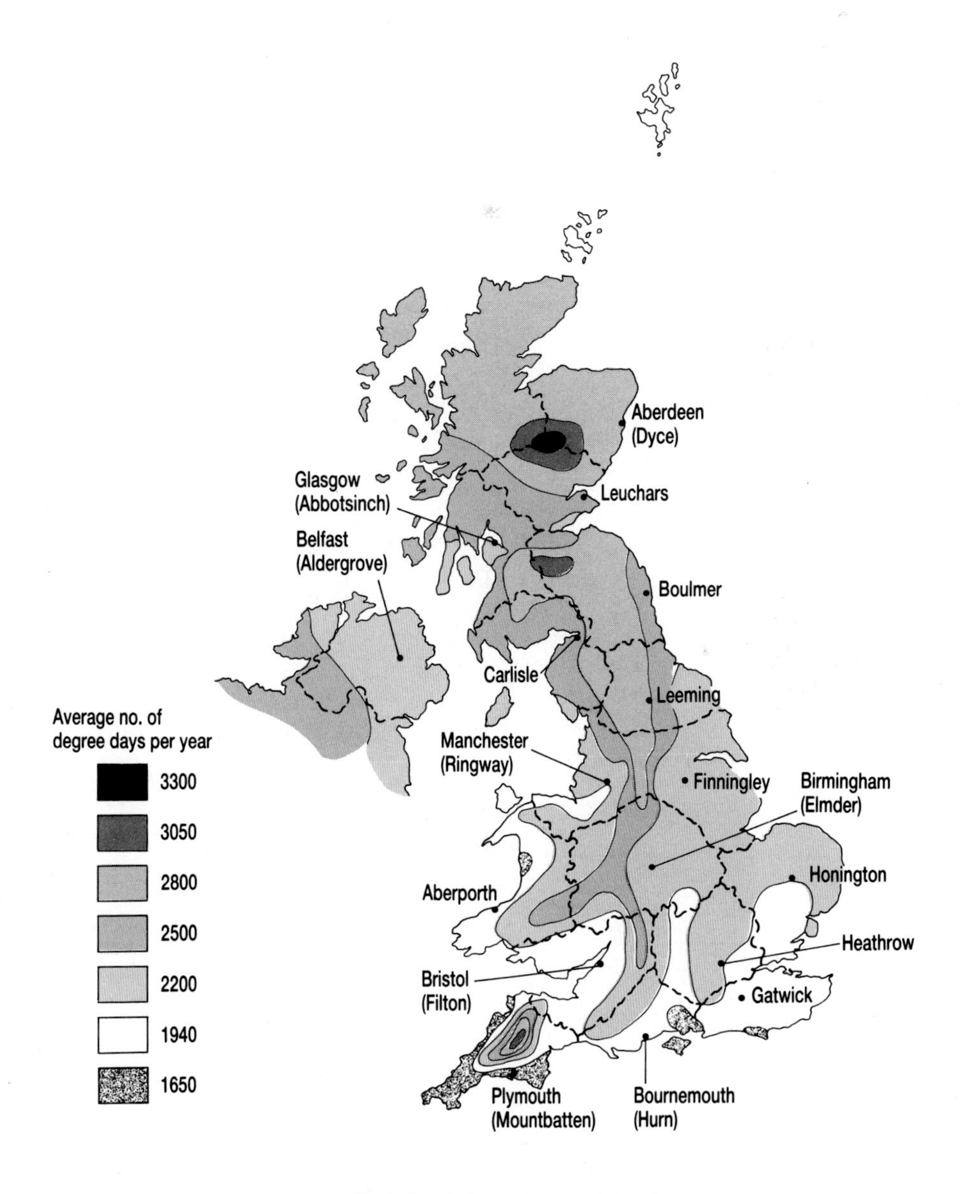

Fig 2.13 A degree day map of the UK.

QUESTION

2.2 A student lives in a bedsit in Plymouth, and pays annual heating bills of £240. If she moves to a similar bedsit in Aberdeen, how much extra can she expect to pay if she is to keep just as warm?

Large-scale heating schemes

We are used to heating our homes with small, domestic heaters. However, this is not necessarily the most efficient way of providing heat. There are two kinds of large-scale systems which are of growing importance in urban areas.

In a **district heating scheme**, a single large boiler house heats water, which is then pumped around factories, offices and houses in the neighbourhood. Such schemes often involve making use of energy released when refuse is incinerated. Fig 2.14 gives some details of the Nottingham scheme, which provides heat equivalent to that needed for 17 000 homes.

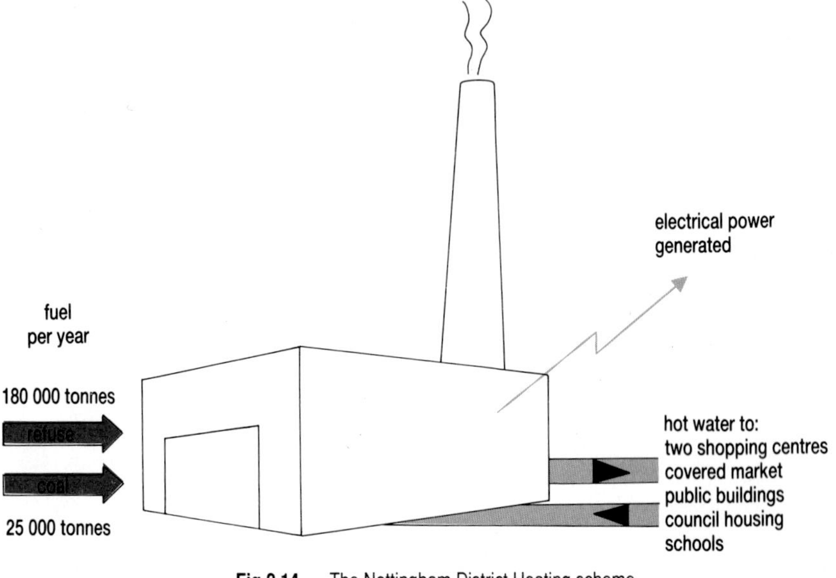

Fig 2.14 The Nottingham District Heating scheme.

In a **combined heat and power (CHP) scheme**, an electricity generating plant is designed to make better use of the waste heat which is otherwise lost. Fig 2.15 shows that, although the amount of electricity generated is less than in a conventional power station, the overall efficiency of the plant is more than doubled. Such a scheme operates in Sheffield, where it supplies heat to 18 multi-storey blocks of flats, as well as shops, public houses and local authority buildings.

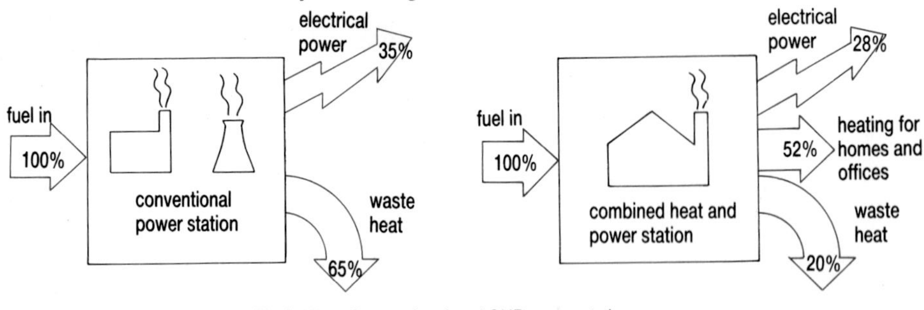

Fig 2.15 Conventional and CHP power stations.

District heating and CHP schemes are proving to be of great importance, and are used extensively in northern Europe. For example, in Denmark more than 25% of heating requirements are currently met by CHP schemes.

QUESTION

2.3 CHP and district heating schemes are becoming increasingly popular. List some advantages and disadvantages of such schemes, compared to heating systems in individual homes. Think about efficiency, environmental impact, employment, convenience and controllability.

ASSIGNMENT

Living with building regulations

The design of buildings and the choice of construction materials are controlled by building regulations. These lay down minimum standards for materials, and the maximum permitted area of single-glazed windows. Table 2.3 shows the regulations which came into effect in 1990.

Table 2.3 Thermal insulation regulations

Building type	Maximum U-values of elements/Wm^{-2} K^{-1}			Maximum single-glazed areas	
	floors	walls	roofs	windows	rooflights
dwellings	0.45	0.45	0.25	15%	15%
offices, shops	0.45	0.45	0.45	35%	20%
industrial, storage	0.45	0.45	0.45	15%	20%

(Window area as percentage of wall area; rooflight area as percentage of roof area)

Fig 2.16 shows two typical buildings. Assume that they are built to the minimum standards laid down in the regulations.

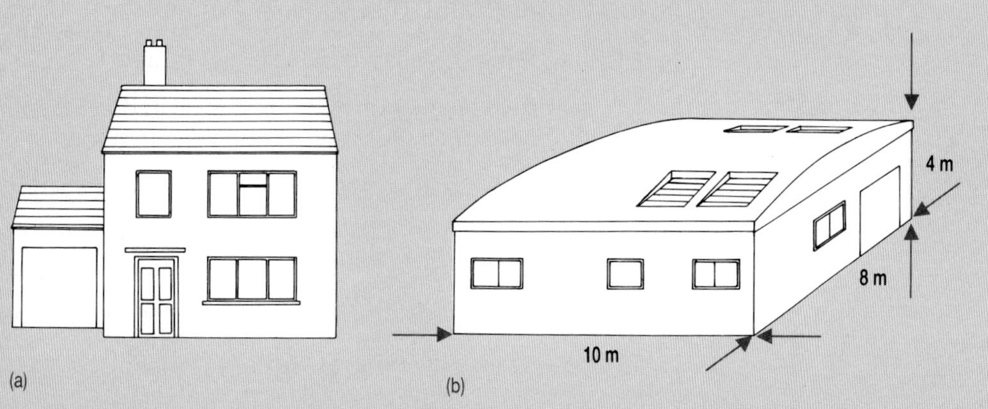

Fig 2.16 Two buildings: **(a)** a house, and **(b)** a warehouse.

1. The house has a living room with an external wall of area 9.2 m^2. The single-glazed window (U = 5.3 W m^{-2} K^{-1}) is 15% of this area. Calculate the rate of heat loss through this wall on a day when the outside temperature is 10 °C less than the inside temperature.
2. The warehouse has the maximum permitted area of single-glazed window and rooflight.
   - **(a)** Estimate the power requirements of a heating system on a day when the temperature difference is 10 °C.
   - **(b)** How much energy would be saved in a day if the windows and rooflights were double-glazed (U = 3.0 W m^{-2} K^{-1})?

SUMMARY

When choosing an energy supply, we often pay for convenience. Electricity is a very convenient form of energy, as it is easily transmitted and easily converted to other forms. The electricity supply industry has to cope with great variations in demand.

As energy conservation becomes more important, new building designs and energy supply systems are coming into use. U-values are a simple way of calculating energy losses by heat conduction from a building.

Chapter 3

RESERVES AND RESOURCES

In this chapter, we will look at patterns of energy consumption, and how these vary historically and geographically. We will look at the resources available to meet human demands, and consider the part which technological developments have to play in making use of limited resources. We will also think about the associated environmental problems, and their possible solutions.

LEARNING OBJECTIVES

After studying this chapter you should be able to:

1. interpret information about historical and geographical differences in energy supply and consumption;
2. construct and interpret Sankey diagrams for energy conversion processes;
3. distinguish between energy reserves and resources;
4. discuss the implications of changes in energy policy;
5. use a variety of energy units;
6. provide scientific descriptions of the environmental impact of different energy sources, and evaluate their possible future outcomes.

3.1 PATTERNS OF CONSUMPTION

The energy sources which humans make use of show great variation, both from place to place and from age to age. Fig 3.1 summarises the historical development of our use of energy sources and energy technologies.

Particular choices of energy sources are determined by a variety of factors:

- the availability of fuel;
- the availability of technology;
- the state of industrialisation.

For example, an industrial society makes less use of wood as a fuel than an agricultural society. Nuclear technology has only been available since the 1950s, since when its importance has grown rapidly.

Fig 3.2 shows how annual energy consumption per person varies from country to country. This makes clear the great difference in energy consumption between a 'consumer' society such as the USA and a more rural, village-based society such as India.

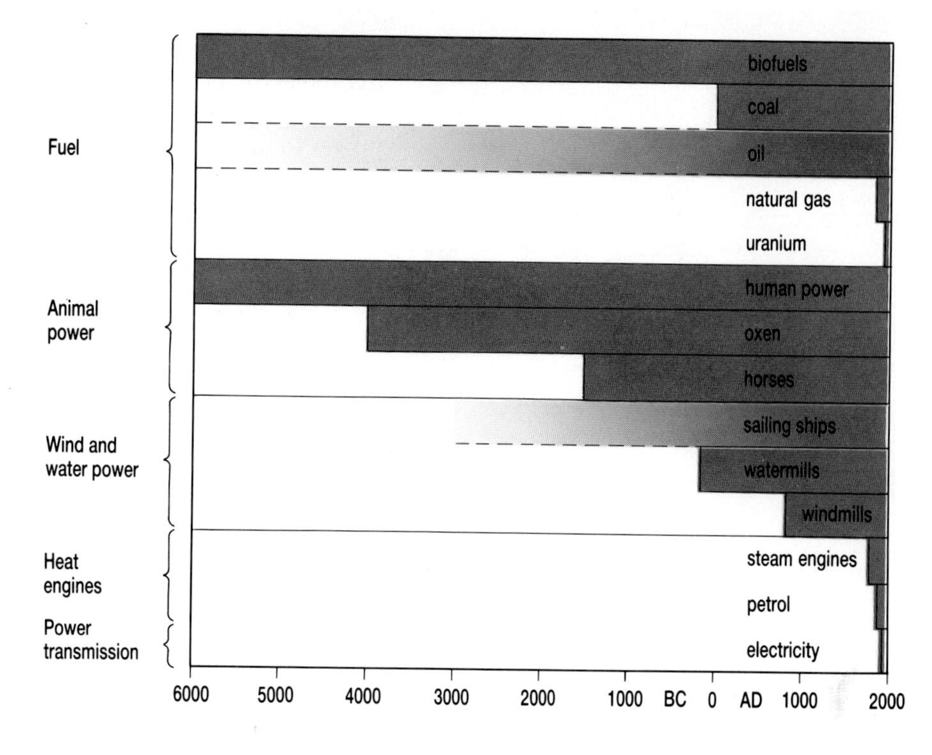

Fig 3.1 Energy technologies have become important at different times in history.

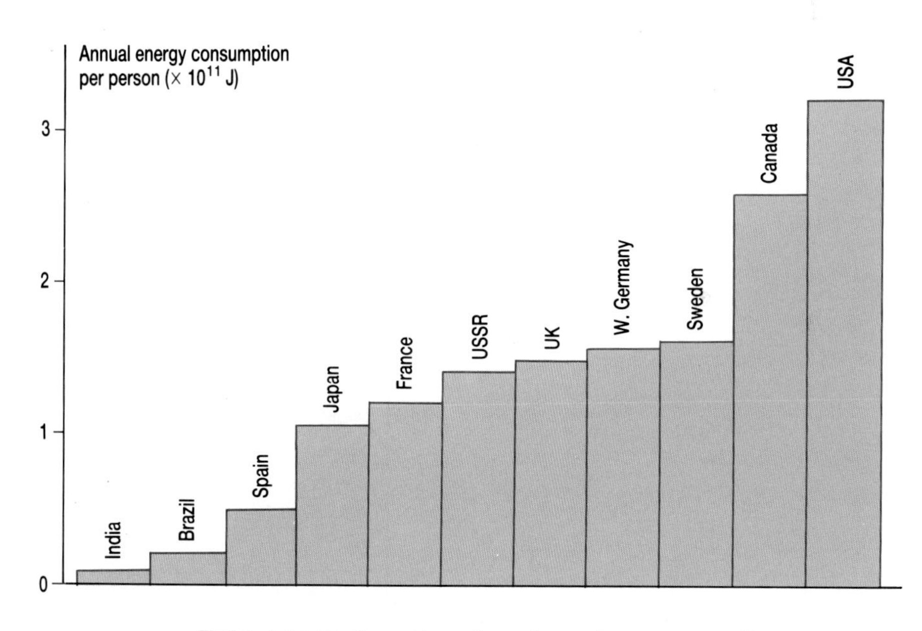

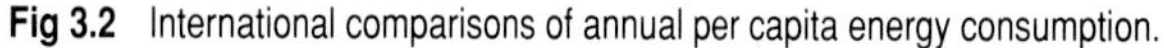

Fig 3.2 International comparisons of annual per capita energy consumption.

ASSIGNMENT

Energy trends

In Fig 3.3, you can see how different primary energy sources have contributed to total consumption in the UK and the USA since 1850. Study these graphs, and answer the questions which follow.

1. Both graphs show a decline in the use of wood as a fuel. Why do you think this happened later in the USA than in the UK?
2. What reasons can you suggest for the decline in the importance of coal as a fuel? Why has oil become more important?

3. On the basis of these graphs, would it be safe to conclude that hydroelectric power has reached the limit of its contribution to energy supplies in these two countries? Can we expect nuclear power to increase in importance?

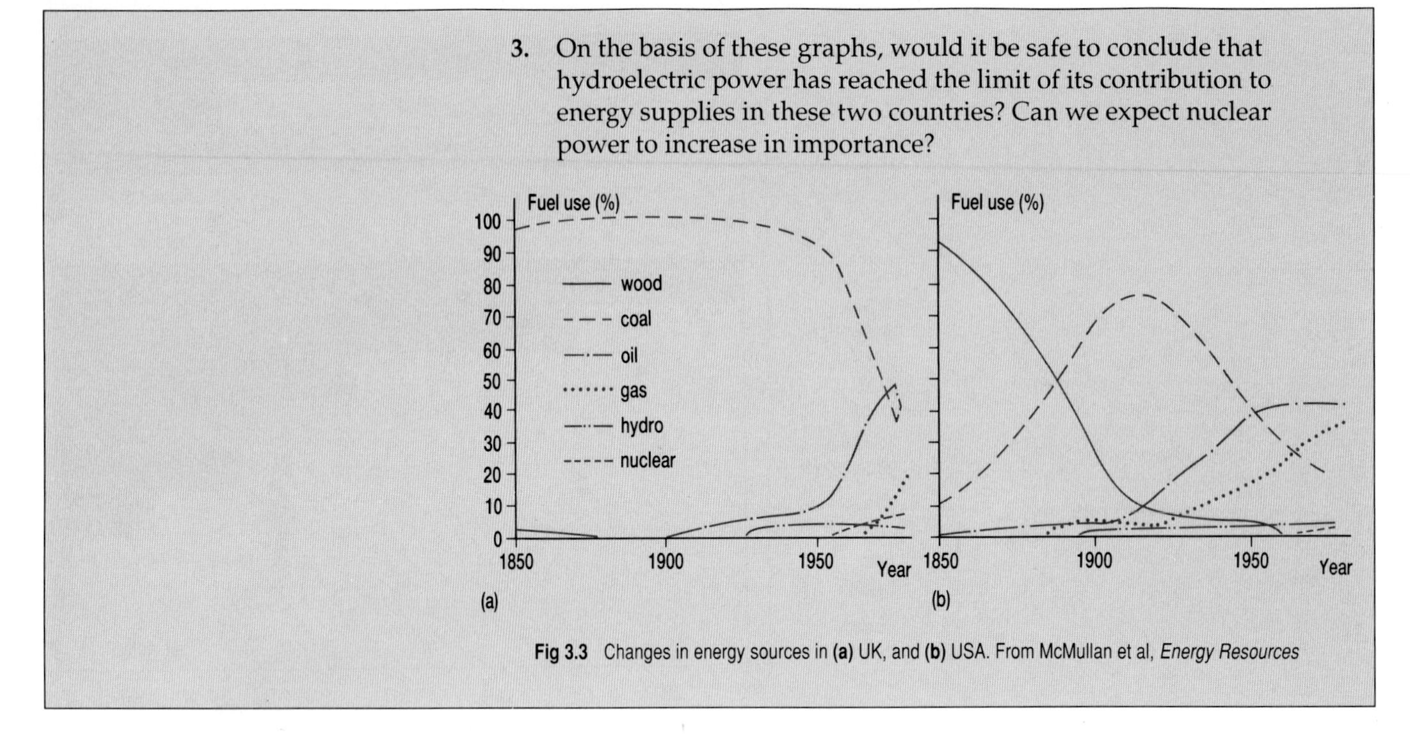

Fig 3.3 Changes in energy sources in **(a)** UK, and **(b)** USA. From McMullan et al, *Energy Resources*

3.2 UK ENERGY FLOWS

To develop a more detailed picture of the way in which energy uses are changing, and how they may change in the future, we will concentrate on the picture in the United Kingdom. Fig 3.4 shows changes in UK primary energy demand since 1950. This is thus a closer look at part of Fig 3.3(a) – but note that this graph shows actual energy consumption, rather than percentages of the whole.

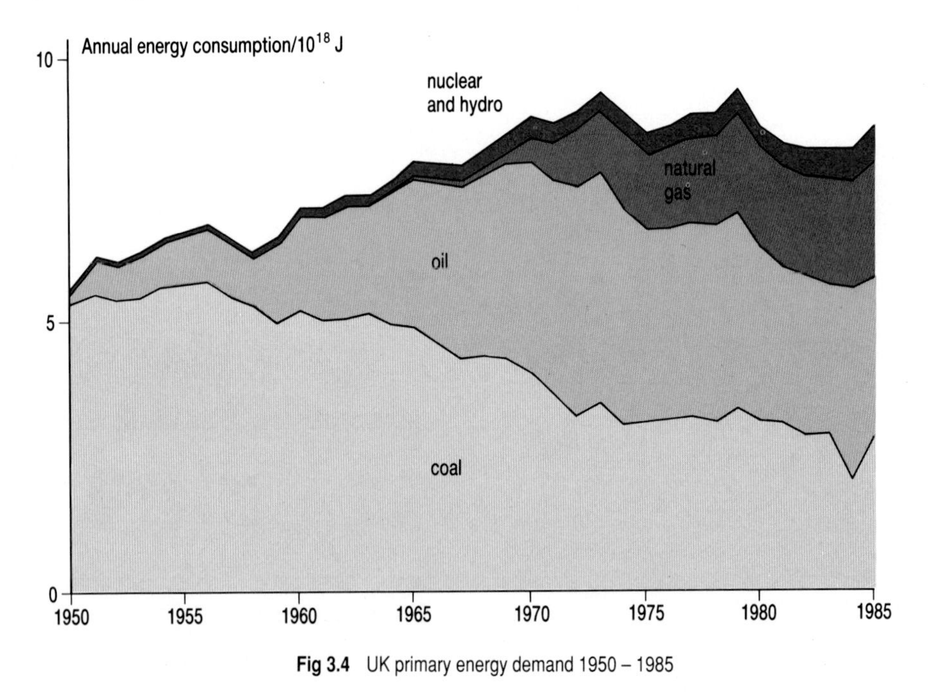

Fig 3.4 UK primary energy demand 1950 – 1985

Fig 3.4 shows the demands made on **primary energy sources**. How effectively is this energy used? All energy sources must be converted into more useful forms before use. This inevitably results in losses. In fact, we make good use of just about 40% of the energy taken from primary sources. Fig 3.5 shows what becomes of the energy between its source and the end-user.

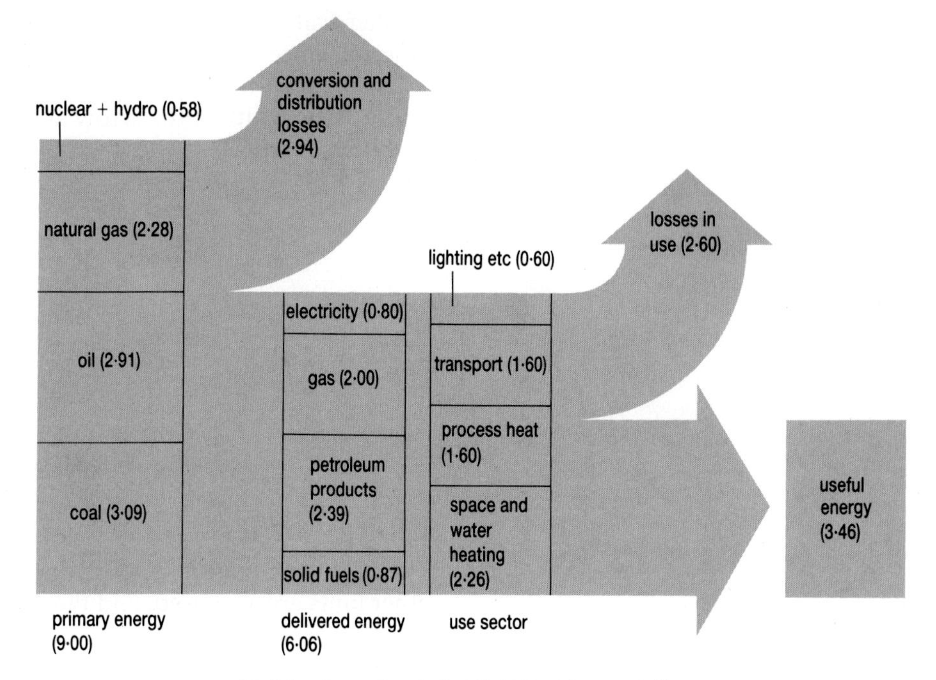

Fig 3.5 Energy flows in the UK, 1987; all figures × 10^{18} J.

Of course, all energy conversion processes are less than 100% efficient. For example, an electric filament lamp is only about 5% efficient as a producer of light; the remaining electrical energy is converted to heat. The laws of thermodynamics (see Chapter 4) dictate that any fossil-fuel power station cannot be much more than 40% efficient in producing electrical energy.

The point to appreciate here is that some losses in energy conversion processes are inevitable, while others can be avoided. Whether an overall loss of 60% is acceptable is another question.

QUESTIONS

3.1 Study Fig 3.5. Explain what you understand by the terms 'primary energy', 'delivered energy' and 'useful energy'.

3.2 Which is the largest use sector, and which is the smallest?

3.3 Consider the following processes:
(i) Natural gas is extracted from a North Sea well, and is eventually used to boil a kettle of water on a gas stove.
(ii) Coal is mined for use in a power station; the electricity is used to boil water in an electric kettle.
(a) For each process, identify the energy changes which occur. Where do energy losses take place?
(b) Which process do you think is the less efficient overall?
(c) Which would you expect to be the cheaper way of boiling water for a cup of tea?

Sankey diagrams

Fig 3.5 is an example of a Sankey diagram. This is a way of representing energy changes, and shows two things. Firstly, for a particular process, it shows the changes which take place between different forms of energy, and the final forms that the energy takes at the end of the process. Secondly, the width of the arrow shows the proportions of energy in different forms at each stage of the process. Fig 3.5 is a Sankey diagram for the whole 'energy economy' of the United Kingdom. Fig 3.6 shows a Sankey diagram for a smaller-scale process: a pulley is used to lift a weight. The process is 90% efficient, since of the 200 J of work put into the system, 20 J are lost as heat.

Note that a Sankey diagram reflects the Principle of Conservation of Energy. The width of the arrow at the start is equal to the sum of the widths of the final arrows.

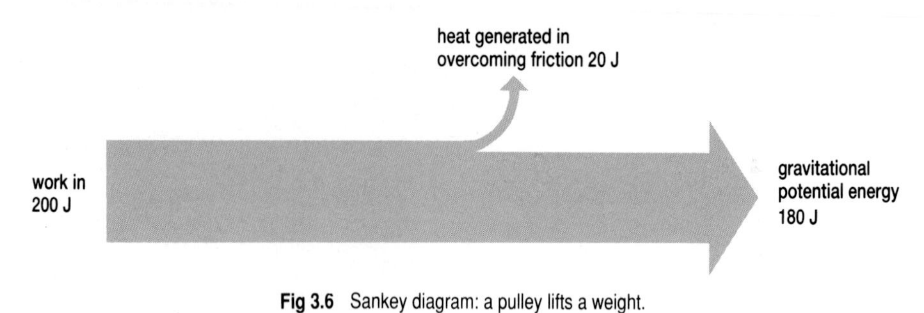

Fig 3.6 Sankey diagram: a pulley lifts a weight.

Energy units

The SI unit of energy is, of course, the joule. However, a great variety of other units are also used, and these are explained in Table 3.1.

Table 3.1 Energy units

Unit	Used for	Conversion
kilowatt-hour (kW h)	electricity	1kW h = 3.6×10^6 J
therm	gas	1 therm = 1.055×10^8 J
calorie	heat energy	1 calorie = 4.186 J
tonne of coal equivalent (tce)	fossil fuel	1 tce = 26.6×10^9 J
tonne of oil equivalent (toe)	fossil fuel	1 toe = 1.7 tce = 45.2×10^9 J
barrel	oil	1 barrel = 5.6×10^9 J
cubic metres (m^3)	gas	1000 m^3 = 1.43 tce = 38.0×10^9 J
kilogram	uranium	5.8×10^{11} J

QUESTIONS

3.4 Fig 3.7 shows the energy changes which take place in a torch in 1 s. Use the diagram to calculate the efficiency with which the torch converts chemical energy stored in the battery into light.

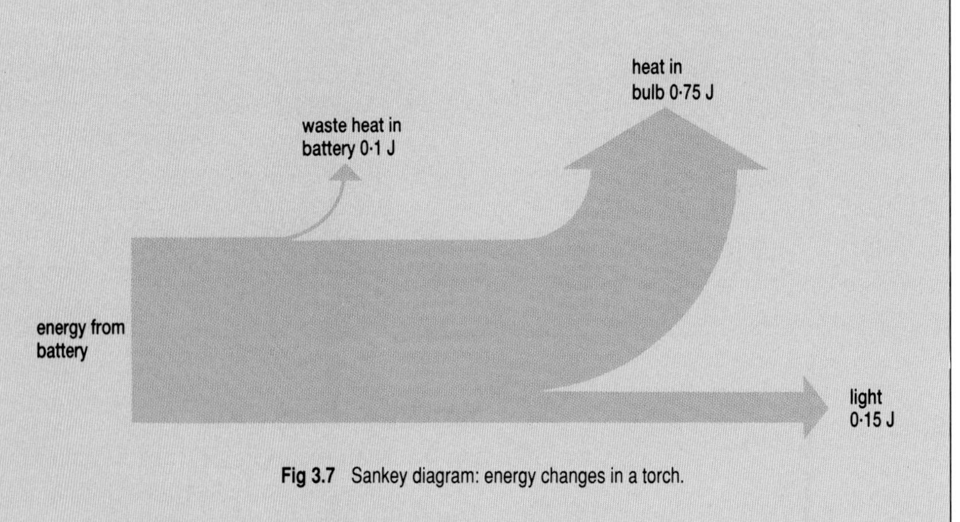

Fig 3.7 Sankey diagram: energy changes in a torch.

3.5 Table 3.2 shows the different forms of energy produced by a combined heat and power (CHP) station in 1 s. Use this data to draw a Sankey diagram for the CHP station.

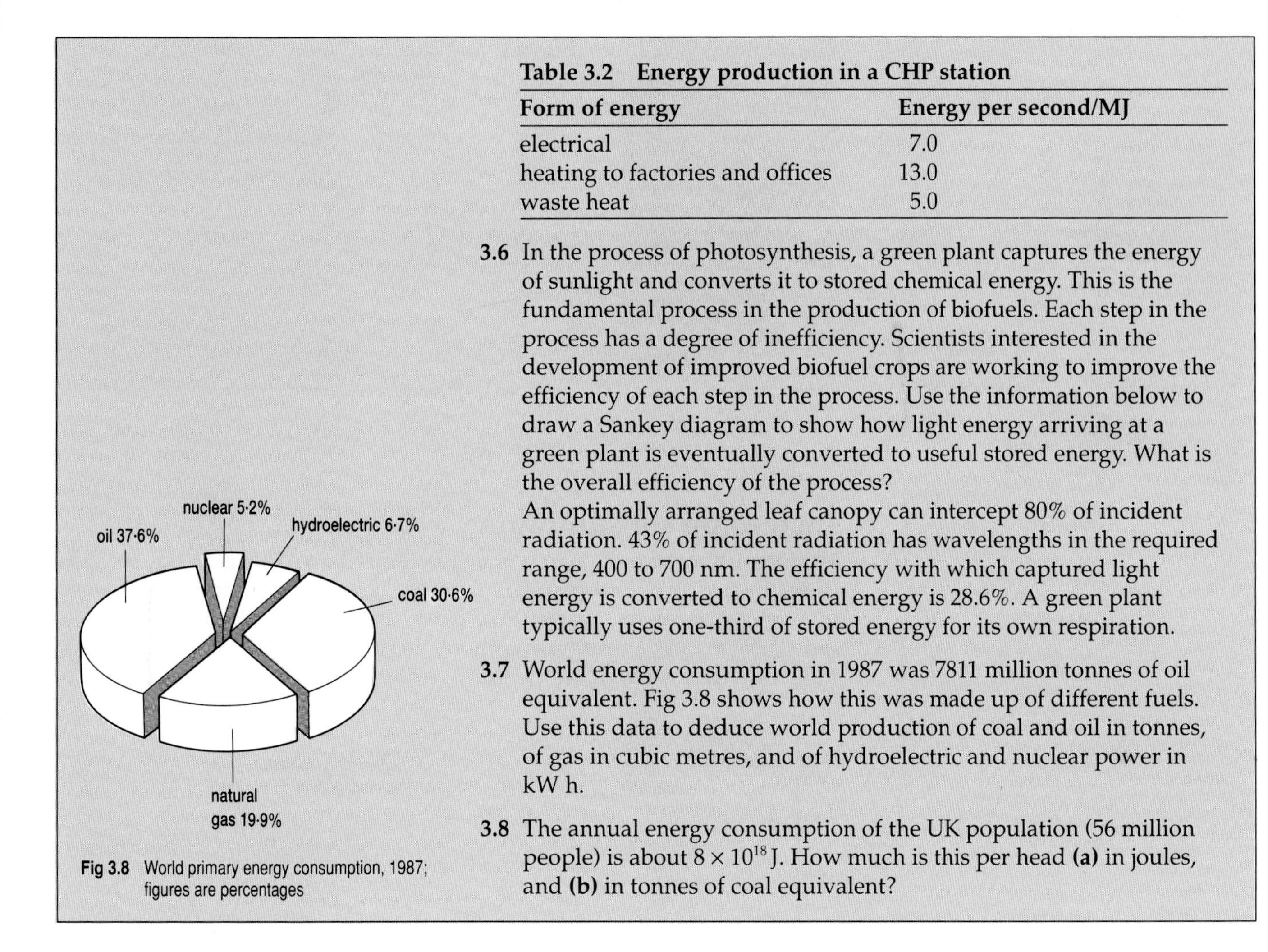

Fig 3.8 World primary energy consumption, 1987; figures are percentages

Table 3.2 Energy production in a CHP station

Form of energy	Energy per second/MJ
electrical	7.0
heating to factories and offices	13.0
waste heat	5.0

3.6 In the process of photosynthesis, a green plant captures the energy of sunlight and converts it to stored chemical energy. This is the fundamental process in the production of biofuels. Each step in the process has a degree of inefficiency. Scientists interested in the development of improved biofuel crops are working to improve the efficiency of each step in the process. Use the information below to draw a Sankey diagram to show how light energy arriving at a green plant is eventually converted to useful stored energy. What is the overall efficiency of the process?
An optimally arranged leaf canopy can intercept 80% of incident radiation. 43% of incident radiation has wavelengths in the required range, 400 to 700 nm. The efficiency with which captured light energy is converted to chemical energy is 28.6%. A green plant typically uses one-third of stored energy for its own respiration.

3.7 World energy consumption in 1987 was 7811 million tonnes of oil equivalent. Fig 3.8 shows how this was made up of different fuels. Use this data to deduce world production of coal and oil in tonnes, of gas in cubic metres, and of hydroelectric and nuclear power in kW h.

3.8 The annual energy consumption of the UK population (56 million people) is about 8×10^{18} J. How much is this per head **(a)** in joules, and **(b)** in tonnes of coal equivalent?

3.3 RESERVES OR RESOURCES?

More than 90% of the energy we use in the world today derives from fossil fuels – coal, oil and gas. These are necessarily limited in their total amounts available to us, and in this section we will look at the future prospects for these energy supplies.

A distinction has to be drawn between the terms reserves and resources. **Resources** include all known quantities of a particular fuel; **reserves** refers specifically to those quantities which it would be worth extracting commercially at today's prices, using today's technology. Thus reserves are a fraction of resources.

Table 3.3 shows estimates of world fuel reserves. Such estimates are inevitably very uncertain, and should be treated with caution. Don't be surprised if you see rather contradictory figures elsewhere.

Table 3.3 Estimates of the Earth's fuel resources

Fuel	Known reserves $\times 10^{21}$ J	Additional resources $\times 10^{21}$ J
coal and peat	24.7	53.6
oil	2.0	27.7
natural gas	0.7	7.6
oil shale, tar sands	0	13.1
total	27.4	102.0

Since the designation of reserves involves deciding whether their extraction is a commercial proposition, it follows that the amount of reserves of a particular fuel can fluctuate with changes in the world market.

For example, it is more difficult, and therefore more expensive, to extract oil from the stormy North Sea than from wells in the Middle East or USA. After the oil crisis of the early 1970s, when the price of oil rose dramatically, North Sea oil suddenly became economically viable, and world reserves of oil rose correspondingly. The resources of oil did not change, but the reserves did.

QUESTIONS

3.9 Explain how the known resources of a particular fuel might increase, and how they might decrease.

3.10 Explain why the reserves of oil shale and tar sand shown in Table 3.3 are zero. Under what circumstances might this figure increase?

Increasing consumption

World consumption of energy is increasing rapidly. This is partly because of the increasing world population, partly because of the industrialisation of more parts of the world, and partly because people in the existing industrialised nations are using more energy year by year. Fig 3.9 shows how this increase has developed since 1850.

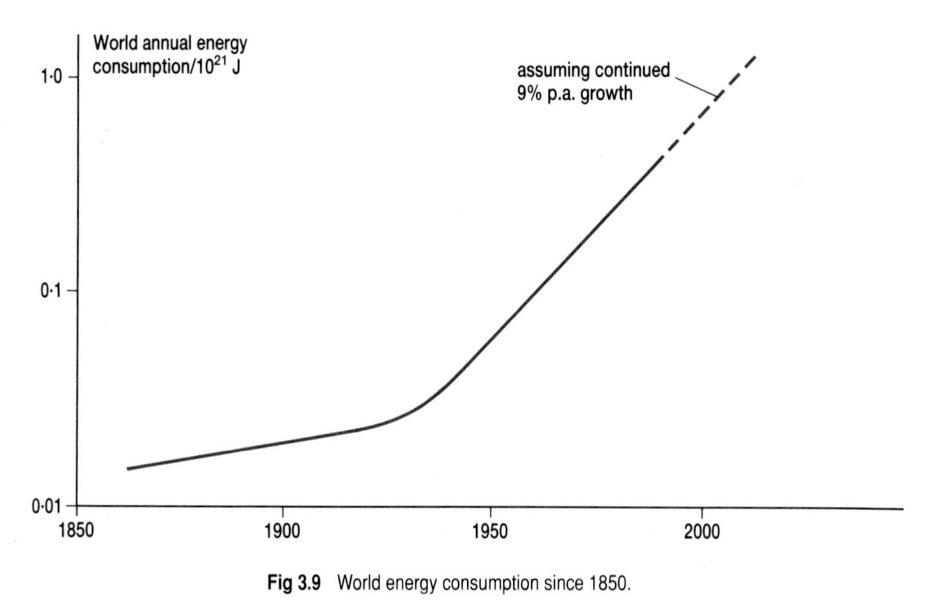

Fig 3.9 World energy consumption since 1850.

Note that the energy consumption scale of this graph is logarithmic. A straight line on the graph thus represents exponential growth in energy consumption. There was a clear increase in the rate at which energy consumption was growing in the middle years of the twentieth century; you might be able to think of reasons for this.

For how long can existing fuel reserves supply our needs if we continue to follow our present pattern of energy consumption? Of course, all resources are finite, but will our fuel resources last for a million years, a thousand years, or will they run out in our lifetimes? Is there an urgent problem?

Fig 3.10 shows how our cumulative consumption of energy is increasing. The three sloping lines show how soon consumption will have overtaken available reserves, assuming steady consumption, and 5% and 10% annual growth rates. It is not difficult to see that, for the growth rate shown in Fig 3.9, consumption will outstrip reserves in no more than two generations. Even consumption with no growth will eventually bring problems.

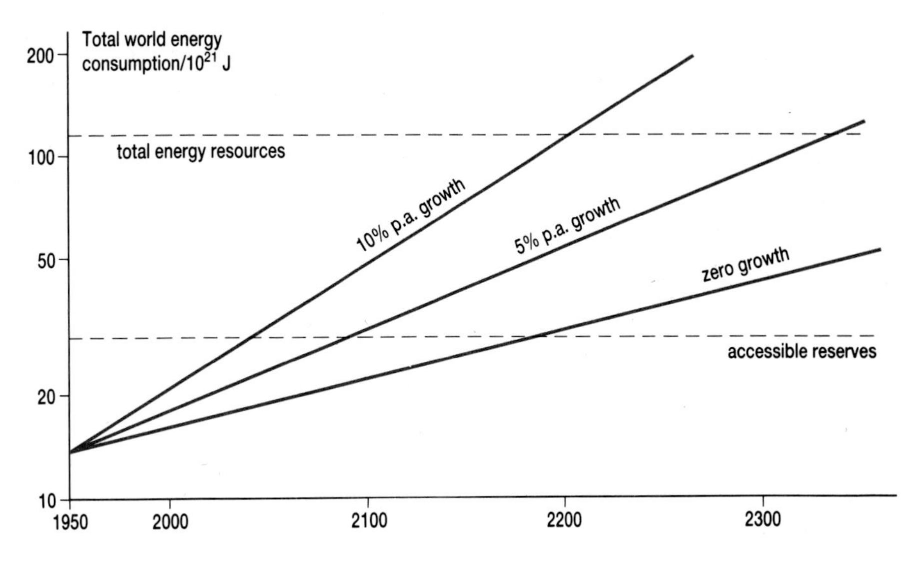

Fig 3.10 Cumulative energy consumption.

A word of caution

Energy forecasting is a notoriously tricky business. Prospects change from year to year, and estimates of reserves can only be approximate. At the end of the 1960s, oil was cheap, and gas was burnt off as a wasteful by-product – Fig 3.11. Then came an oil crisis, and prices rose sharply. The potential of gas was realised, and predictions of a great expansion of nuclear power were being made. However, new reserves of oil were discovered faster than existing reserves were being used up, and oil prices collapsed again. By the end of the 1980s, the world seemed awash with oil and gas, and the nuclear industry was in recession.

What does seem likely, however, is that, for a variety of economic and environmental reasons, we will be encouraged to take energy-saving and renewable energy sources more seriously in the future.

Fig 3.11 Gas flares burn above a Middle East oil well. This wasteful and polluting practice is now done less than formerly.

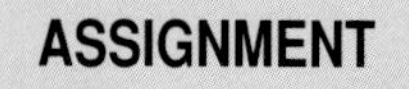

ASSIGNMENT

A low energy strategy for the UK

The extract which follows is taken from *Energy without End* by Michael Flood, published by Friends of the Earth. It advocates a radical change in the UK Government's energy strategy. Your task is to think about the policy advocated, and to consider its implications. Write an assessment of the consequences of this strategy.

You should think about the following points:

(a) How will the energy supply industry be affected?
(b) What are the implications for building design, for industry and for transport?
(c) How might your everyday life be affected?

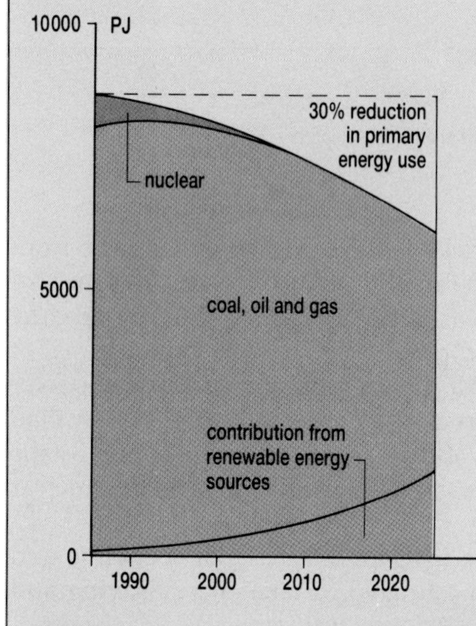

Fig 3.12

A Future Worth Working For

The Government's energy policy clearly fails to address the twin problems of gross inefficiency in the supply and use of energy in the UK and the provision of clean, safe and secure sources of energy for the future. A new approach is urgently needed.

The strategy described below has three main components:

- a determined drive to stamp out the inefficient use of energy in all sectors of the economy;
- the introduction of cleaner fossil fuel-burning technologies and the widespread adoption of combined heat and power; and
- a concerted programme of investment in renewable energy technologies.

The **minimum** targets should be:

- an average reduction in primary energy use of 1% per annum over the next 40–50 years; and
- an increase in energy generated from renewables to at least 20% of primary energy over the same period.

Some estimates suggest that the UK already gets over 2% of its energy from renewables, principally hydro-power and biofuels, and unaudited solar gains in buildings. This is actually **one seventh** of the proposed 2025 contribution.

Improving Efficiency

It is not necessary to increase energy use in order to improve standards of living. Indeed, the opposite is the case. Improved energy efficiency increases economic growth and raises living standards. Dozens of detailed studies have shown just how much energy could be saved if the Government were to pursue a vigorous nationwide campaign to discourage energy wasteful habits and ensure that fossil fuels were used more efficiently.

Technically, it would be possible to reduce primary energy demand by **two thirds** or more through a combination of efficiency improvements and a switch away from using premium fuels (such as oil and gas) and electricity to supply low temperature heat. The kind of measures that would be required are comparatively simple and straightforward.

3.4 ENERGY PROBLEMS AND SOLUTIONS

We have seen that there is a major problem for the future which arises from the growth in energy demand, which cannot be met by the world's finite resources. But there are other problems which arise from the way in which we use energy. In this section, we will take a look at two of these problems.

The greenhouse effect

The temperature of the Earth's surface is rising; Fig 3.13 shows how it has changed since 1960, and how it is predicted to rise in the future. Such temperature changes could have a drastic effect on human life. During the 'mini Ice Age' of the sixteenth and seventeenth centuries, when temperatures were only a degree or two lower than in the twentieth century, the Thames frequently froze in London. More seriously, the Indians of central North America were forced to abandon their settled agricultural way of life and become nomadic hunters, as their traditional crops could no longer grow. The climatic changes we are currently witnessing may have equally serious consequences.

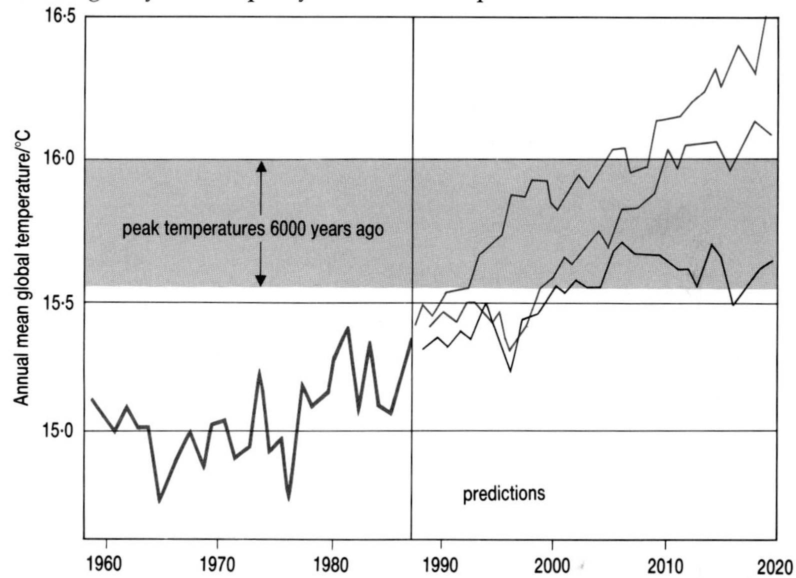

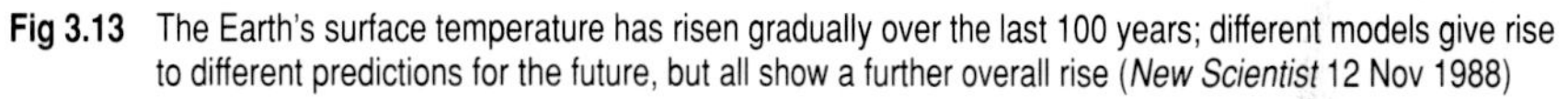
Fig 3.13 The Earth's surface temperature has risen gradually over the last 100 years; different models give rise to different predictions for the future, but all show a further overall rise (*New Scientist* 12 Nov 1988)

Temperature increases appear to be linked to the increase in carbon dioxide in the atmosphere. Fig 3.14 shows an observed rise of 3 or 4% per

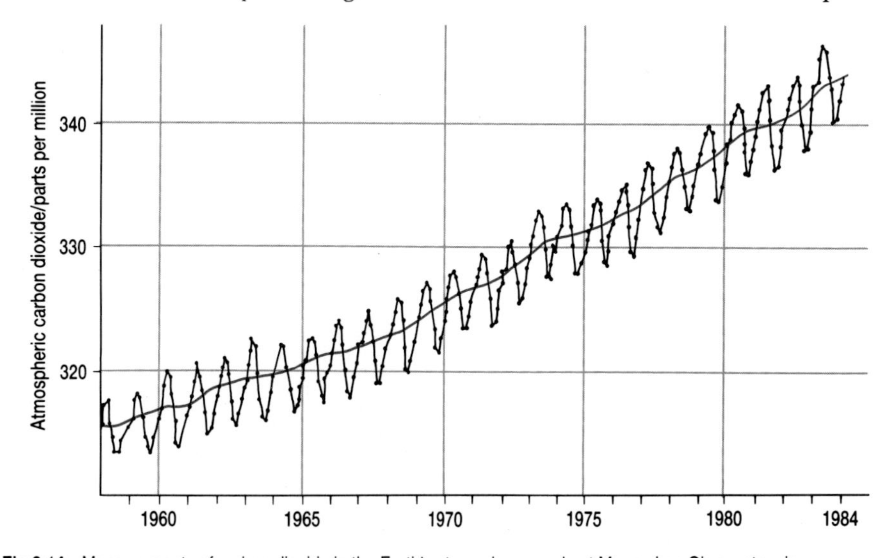

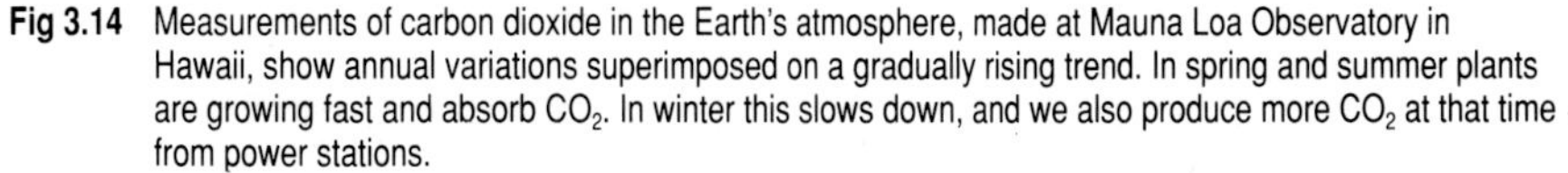
Fig 3.14 Measurements of carbon dioxide in the Earth's atmosphere, made at Mauna Loa Observatory in Hawaii, show annual variations superimposed on a gradually rising trend. In spring and summer plants are growing fast and absorb CO_2. In winter this slows down, and we also produce more CO_2 at that time from power stations.

decade in the CO_2 concentration. As the effect accelerates, we may see a doubling of CO_2 concentration in the next century. Most of this increase has come about because of our burning of fossil fuels.

Carbon dioxide is only one of the gases which is thought to contribute to the warming shown in Fig 3.13. Methane, nitrous oxide, chlorofluorocarbons and ozone also contribute, and all are increasing due to human activity.

How does this warming come about? To understand this, we need to think about why the Earth is warmer than the Moon. Both Earth and Moon are at the same average distance from the Sun, and hence might be expected to have the same average temperature. However, the Moon is more than 30 °C colder on average than the Earth. Why the difference?

The Earth is kept warm by an insulating blanket, its atmosphere. The Moon, of course, has no atmosphere. The way in which the atmosphere helps to maintain a warm surface temperature is shown in Fig 3.15; this is usually likened to the way in which a greenhouse stays warm relative to its surroundings on a sunny day.

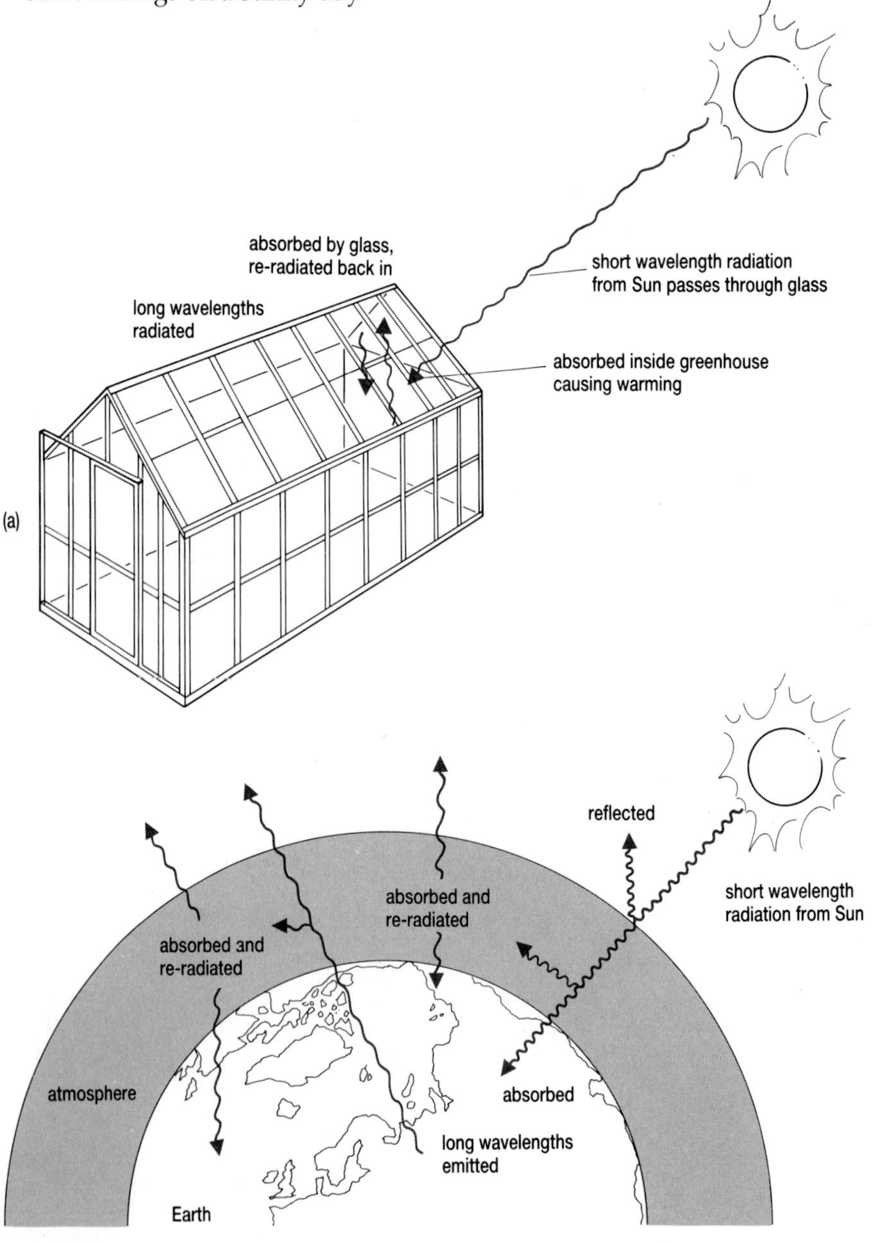

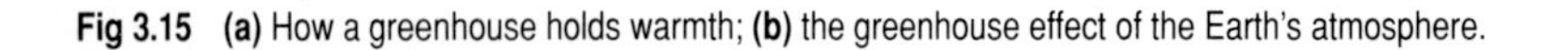

Fig 3.15 **(a)** How a greenhouse holds warmth; **(b)** the greenhouse effect of the Earth's atmosphere.

Radiation from the Sun reaches the atmosphere; some is reflected, some is absorbed, and some passes through to the Earth's surface. The radiation which arrives at the Earth's surface is largely in the visible and infra-red regions of the spectrum. Subsequently, the Earth's surface re-radiates this absorbed energy as infra-red. Because the greenhouse gases, and water vapour in the atmosphere, preferentially absorb infra-red radiation, much of the re-radiated energy is absorbed by the atmosphere and then re-radiated back to the Earth's surface. This serves to maintain an average temperature of about 15 °C. An equilibrium is established, with the infra-red radiation radiated by the Earth balancing the visible and infra-red radiation arriving from the Sun.

The effect of increasing CO_2 concentration is to upset this balance. The atmosphere absorbs more infra-red; to restore equilibrium, the Earth's temperature must rise by a few degrees.

ASSIGNMENT

Living in a greenhouse

What might the consequences of a rise in global temperatures be? Can we be sure that temperatures will continue to rise? It is important to realise that our scientific understanding of the Earth's weather system is constantly changing. In the 1970s, climatologists warned of the possible imminent onset of a new Ice Age. Perhaps the increase in CO_2 concentration will be reversed by increased activity of marine micro-organisms. Perhaps there will be benefits from global warming – increased crops in temperate regions, increased rainfall, less demand for energy for space-heating – which will outweigh the negative effects.

It is difficult to know what are the true trends in the Earth's climate. One hot summer does not prove that the greenhouse effect is at work; one cold winter does not mean the next Ice Age is upon us. Some scientific knowledge is needed to be able to say what may really be happening.

Newspapers and scientific magazines frequently have articles about trends in the weather. Make a collection of some articles, and try to answer the following points:

1. What is the current scientific picture of changes in the Earth's weather system?
2. What evidence is there to support this picture?
3. How has the scientific picture changed? Do all scientists agree, and what are the areas of uncertainty?
4. Do newspaper articles give a fair representation of the scientific view?

Nuclear waste

One technology which might contribute to a slowing down of the greenhouse effect is nuclear power. Fast reactors, making the most effective use of available uranium, could perhaps satisfy a large proportion of our energy demands. This might reduce our dependence on fossil fuels, which release carbon dioxide into the atmosphere when they are burnt. However, nuclear power (like all technologies) has its negative side. How can we restrict the release of radioactive materials into the environment? In particular, how can we cope with the waste which results from nuclear fission?

There are two general approaches to nuclear waste management: dispersal and containment. **Dispersal** involves drastically reducing the concentration of nuclear materials by dispersing them, usually in the sea.

Containment may involve concentrating the most hazardous waste, and then storing it in a safe place. Because of the long half-lives of many of the isotopes involved, storage may be necessary for a very long time – many human lifetimes.

ASSIGNMENT

Coping with nuclear waste

The nuclear industry provides a lot of information on how it deals with nuclear waste. Its opponents publish detailed critiques. Find some of these publications, and use them to answer the points below.

1. Explain how waste may be classified as low-level, intermediate and high-level.
2. How are these different grades of waste dealt with? Which are dispersed, and which are contained?
3. What criticisms are made of these methods? To what extent do you think these criticisms are supported by scientific evidence?
4. All forms of energy supply have environmental impacts. List those for (i) nuclear power, (ii) fossil fuels, and (iii) renewable resources. Which do you think are likely to be the most serious in the short term, and in the long term?
5. Radioactive materials are easy to detect, even at very low concentrations. (Each click from a Geiger counter tells us that the nucleus of a single atom has decayed.) A single nuclear decay may cause cellular damage, leading to cancer. An opponent of nuclear power might suggest that the nuclear industry is casual in its handling of lethal materials. A supporter of the industry might suggest that its opponents are unnecessarily concerned over trivial risks. How does the industry try to minimise these risks? What evidence is there of damage to the health of workers in the industry, and of the general public?

SUMMARY

Historically, human technology has developed to give us access to an increasing range of energy sources and supplies. Consumption of energy tends to increase with industrialisation.

We are beginning to understand that energy consumption cannot increase in an unlimited way. Reserves are limited, and the burning of fossil fuels may lead to the greenhouse effect with serious problems of global warming. Other energy supplies (renewable and nuclear) have environmental impacts which must also be evaluated before we decide to what extent they should be used.

Theme 1: Examination Questions

T1.1

(a) Distinguish between *renewable* and *non-renewable* energy resources, giving *three* examples of each type.

(b) Discuss, with reference to the various energy transformations which take place, the processes by which energy from the Sun, falling on the Earth and its atmosphere, is converted to electrical energy in a hydro-electric power station.

(c) Water falling through a vertical distance of 200 m provides the input power to a hydro-electric turbine. Calculate the speed of the water entering the turbine. What assumptions have been made? Can they be justified in practice?
Assuming your calculated value for the speed of water, calculate the input power to the turbine if the pipe diameter is 1.5 m. If the efficiency of this turbine–generator is 80%, what is the electrical power output?

(ULSEB 1985)

T1.2

(a) The total reserve of geothermal energy is estimated to be 4.1×10^{25} J. Of this resource, about 2% is believed to be hot enough for use in electricity generation of which only about one fifth is thought to be recoverable by existing technology. The conversion efficiency would be about 2.5%.
Current world primary energy consumption is at a rate of about 2×10^{20} J per year of which about one quarter is used for electricity generation.

(i) Explain the meaning of the terms *primary energy* and *geothermal energy*. Suggest two physical phenomena which act as sources of geothermal energy.

(ii) Demonstrate whether the figures suggest that geothermal energy could meet a significant proportion of the world's demands for electrical energy.

(b) (i) Draw a labelled block diagram for a conventional power station using a fossil fuel. Discuss the need for the steam to be heated to a high temperature. (You should make reference to the second law of thermodynamics.)

(ii) Water from geothermal sources generally has a temperature between 75 °C and 200 °C. It is often rich in minerals, including sulphur. Suggest how this water might be used to reduce the fuel consumption of a conventional power station. Suggest one other commercial use for water from geothermal sources.
Discuss briefly the environmental implications of the use of geothermal energy.

(ULSEB 1989)

T1.3

(a) The U-values of four construction components are given below.

Component	U-value/$\text{Wm}^{-2}\,\text{K}^{-1}$
Single-glazed window	5.6
Double-glazed window	3.2
Uninsulated roof	1.9
Well-insulated roof	0.4

What do you understand by the U-value of a component?
A house has windows of area 24 m^2 and a roof of area 60 m^2.
The occupier heats the house for 3000 hours per year to a temperature which on average is 14 K above that of the air outside. Calculate the

energy lost per year through (i) the single-glazed windows, and (ii) the uninsulated roof, expressing your answers in kW h.

If electricity costs 5.5p per unit, calculate the annual savings the occupier could make by (iii) installing double-glazing, and (iv) insulating the roof.

If double-glazing costs £3000 and roof insulation costs £100, which, if either, of the two energy-saving steps would you advise the occupier to take?

(b) To cope with sudden surges in power demand, gas turbine generators are sometimes used, running permanently, but only switched into the grid when required – perhaps only a few hours each year. A far cheaper proposed solution is to switch off the gas generators and to keep a heavy flywheel spinning at high speed. When extra power is required, the wheel drives a generator and its angular speed drops to about one-third of its initial value in about three minutes. By this time the gas generators can be started up to meet the demand.

(i) Give an example of an event which might cause a major power surge.

(ii) The mass, m, of the proposed wheel is 1.6×10^5 kg and its radius, r, is 1.7 m. It will be given an angular speed, ω, of 240 rad s^{-1}. Calculate its kinetic energy, E_k, where $E_k = \frac{1}{4}mr^2\omega^2$.
Comment on the use of this flywheel to cope with a sudden surge in the electrical power demand.

(iii) Name one other type of power station which is capable of being brought on-line quickly to meet a power surge. Name one type which is not capable of such a fast response. Explain the reasons for the difference in response times.

(ULSEB 1989)

T1.4

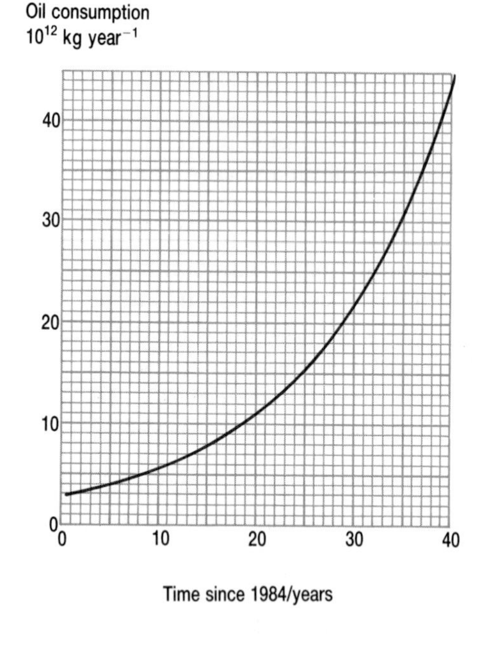

(a) 'Oil is one of the world's fossil fuels.' What do you understand by this statement? In 1984, estimates suggested that the total usable mass of oil in the world was about 300×10^{12} kg. At that time the world's oil consumption was about 3×10^{12} kg per annum and increasing annually by a factor of 107/100, i.e. at an annual rate of 7%, as shown in the graph.

Assuming this rate of increase is maintained, estimate, using the graph or otherwise,

(i) the annual rate of oil consumption in the year 2000, and
(ii) how long this energy resource will last.

For (ii) explain carefully your method of working.

(b) In a nuclear power station a thermal fission reactor is used as the boiler for the production of steam. Three essential constituents of such a reactor are the *fuel*, the *moderator* and the *coolant*. Explain the meanings of the three words in italics and the role each of these constituents plays in the working of the reactor.
Explain why a failure of the coolant flow could be disastrous for a reactor even if the control rods shut-down system were fully operational.

(ULSEB 1988)

T1.5

(a) 'With few exceptions, mankind derives all its energy ultimately from the Sun.'

(i) Discuss briefly three methods by which energy is derived from the Sun, explaining the role of the Sun in each process. Your choices should be as diverse as possible.

Classify each energy resource you mention as renewable or non-renewable.

(ii) State two energy resources which are independent of solar energy.

(iii) Indicate the approximate percentage of Britain's energy need which is supplied by each resource to which you have referred in parts (i) and (ii).

(b) The solar constant at the top of the Earth's atmosphere is 1.37 kW m^{-2}. Explain the meaning of this statement.
Discuss quantitatively the feasibility of siting a 600 MW solar power station on land, at latitude 51°N, in close proximity to areas of high population density.
(The maximum conversion efficiency from solar energy to electrical energy is about 10%.)

(ULSEB 1988)

Theme 2

THE THERMODYNAMICS OF ENERGY SUPPLIES

In trying to solve the world's problems of energy supply, scientists have greatly increased our understanding of how energy can change from one form to another. Technology develops ever more rapidly, so that at times it seems that almost anything must be possible. Indeed, at times, science has been presented as if it held the answer to all problems.

Of course, this is not the case. But what science has been able to do is to point out the limitations of technology. The study of thermodynamics has been a very powerful tool in our understanding of energy transformations, and in helping us to maximise the benefit we can gain from the great variety of machines which we now use.

In this theme, we look at the laws of thermodynamics, and how they can be applied to a particular general type of machine, heat engines.

These lorries have diesel engines. They are being loaded with fridges and freezers. Both the diesel engines and the refrigerators are examples of heat engines, whose performance is governed by the laws of thermodynamics.

Steam engines were an important development of the industrial revolution. They had a great fascination; this is a French model steam engine of about 1840.

Chapter 4

LAWS OF THERMODYNAMICS

Science is the study of the materials from which everything is made, and the processes of change which these materials undergo. We can describe these changes in terms of energy changes. Thermodynamics is the branch of science which deals with these energy changes.

In this chapter, we will look at the laws of thermodynamics, and see something of what they can tell us about our attempts to make good use of energy. Then, in Chapter 5, we will go on to a more mathematical treatment of some important practical applications of the laws.

LEARNING OBJECTIVES

After studying this chapter you should be able to:

1. explain the meaning of the term 'heat engine';
2. state the Zeroth, First and Second Laws of Thermodynamics, and discuss the implications these have for the operation of practical heat engines.

4.1 THERMODYNAMICS

The word 'thermodynamics' suggests a study of heat in motion. A more general definition would be to say that thermodynamics is the study of processes involving energy changes.

Historically, the study of thermodynamics came about because of engineers' increasing interest in machines which converted energy from one form to another. In particular, they were interested in engines which burned fuel and produced mechanical energy.

In fact, thermodynamics is a very general branch of science. It can be applied equally to steam engines, to the evolution of stars, and to the chemical processes of living cells. If you are studying Chemistry, you may well use some thermodynamic ideas to explain the way in which a chemical reaction will proceed.

In this chapter, we will look at how we can make use of the energy available in hot materials, particularly hot gases. We use exploding gases in car engines; in a power station, hot steam turns the turbine to drive a generator. Thermodynamics can tell us a lot about these processes.

You already know a lot about thermodynamics. For example, you know that 'energy is conserved'. No doubt, you have carried out many calculations using this idea. You also have the idea that, in inefficient energy conversions, heat is generated. This is wasteful, as it is difficult to convert the heat back to more useful forms of energy. Heat is different from other energy forms or processes – it cannot be made such good use of.

The ideas that energy is conserved, and that heat is a degraded form of energy, are fundamental to thermodynamics. In the rest of this chapter, we will refine these ideas, and see how they apply in particular to heat engines.

4.2 HEAT ENGINES

As a science, thermodynamics grew up with the Industrial Revolution. Engineers built engines. Stationary engines were used for pumping, for hauling wagons up inclines, for agricultural purposes, and so on. Later, locomotives were built, in which the engine itself moved around on wheels. All of these were heat engines.

A **heat engine** is any device which converts heat energy to useful mechanical energy. A car engine, a steam locomotive, a jet engine and a nuclear power station are all examples of heat engines. (Strictly speaking, a power station contains a heat engine which turns a generator.)

The energy to produce heat usually comes from fossil or nuclear fuel. In principle, you could go in the opposite direction from usual by making a heat engine in which electricity boiled water and the steam turned a turbine.

Heat engines may be divided into two types: internal and external combustion engines.

Internal combustion engine: fuel is ignited in an enclosed space, for example a cylinder, and the explosion pushes a piston. A petrol or diesel engine is an example.

External combustion engine: fuel is burned to raise steam, and the expanding steam turns the blades of a turbine. Alternatively, hot gases from burning fuel may themselves be used to turn a turbine. A gas turbine engine is an example of this.

All heat engines have this in common: energy moves from a hot place to a cold place (from the hot reservoir to the cold reservoir) via the engine, and some of the energy is extracted by the engine as useful mechanical work. Fig 4.1(a) shows this in outline.

Some heat engines are designed to work in reverse: mechanical energy is put in, and heat energy is moved from a cold reservoir to a hot reservoir. A familiar example of a device which does this is a refrigerator. In general, such devices are known as **heat pumps**.

Fig 4.1(b) shows the energy flow of a heat pump.

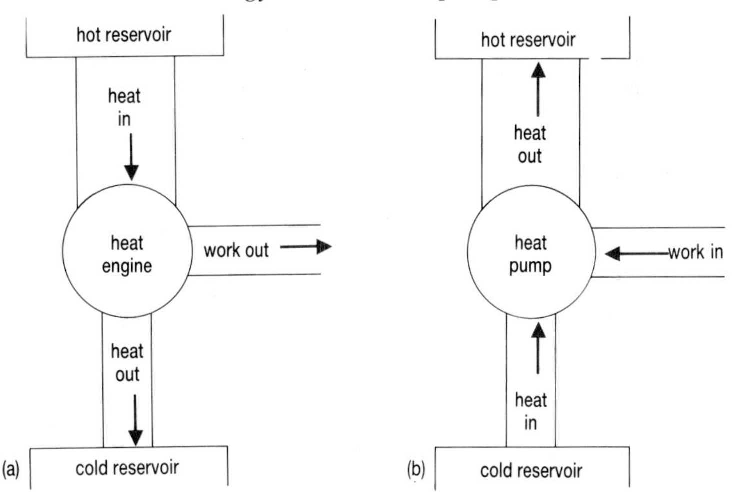

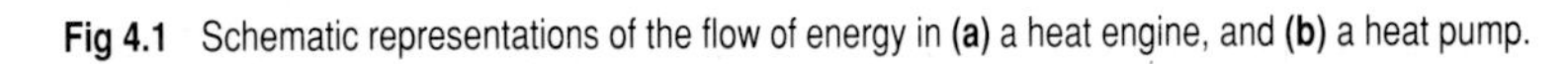
Fig 4.1 Schematic representations of the flow of energy in **(a)** a heat engine, and **(b)** a heat pump.

ASSIGNMENT

Heat engine efficiencies

1. Fig 4.2 shows some energy conversion devices. Which of these are heat engines, and which are heat pumps? Which are neither? Of those which are heat engines, identify those which are internal combustion engines, and those which are external combustion engines.

Fig 4.2 Some energy conversion devices. Which are heat engines? Which are heat pumps?

2. Table 4.1 lists the efficiencies of a variety of energy conversion devices. Your task is to copy out the table, but with the devices regrouped according to the energy conversions which they perform. Group together those which convert fuel (chemical energy) to heat, those which convert between mechanical and electrical energy, and those which convert heat to mechanical energy (heat engines).

Table 4.1 Energy conversion efficiencies

Device	Efficiency
power station boiler	90%
hydroelectric turbine	90%
large electric motor	90%
large electric generator	90%
domestic gas-fired boiler	75%
washing machine motor	70%
domestic coal-fired boiler	60%
steam turbine (power station)	45%
diesel engine	40%
car (petrol) engine	30%
steam locomotive	10%

3. Comment on the typical efficiencies of heat engines.

4.3 THE ZEROTH LAW

It may seem odd to have a zeroth law of thermodynamics. The reason for this curious name is that it was thought up after the first and second laws, but it was recognised to be even more fundamental.

To understand the laws of thermodynamics, it is necessary to have a mental picture of the nature of matter. We will concentrate on gases, but the same laws may be applied equally to solids and liquids.

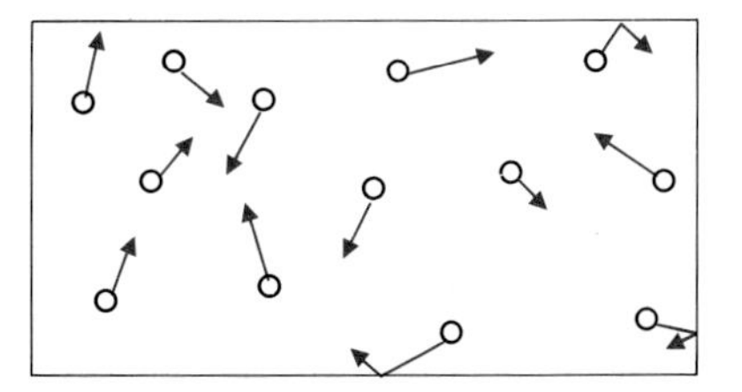

Fig 4.3 The molecules of a gas move randomly, with a range of speeds.

The particles – atoms or molecules – which make up a gas are in constant motion – Fig 4.3. They speed about, colliding with each other and with the walls of their container. The hotter the gas is, the faster its particles are moving. In fact, temperature is simply a way of describing the average energy possessed by the particles.

We have to think about the average energy of the molecules, as some are moving faster than others. As they move about and collide with one another, there is a continual sharing and exchanging of kinetic energy between the particles. A particle may be moving faster at one moment, slower the next. If the temperature of the gas remains constant, however, then so does the average energy of the particles.

The average kinetic energy of the particles (a microscopic quantity) is related to the absolute temperature T (a macroscopic property) by the equation

$$\text{average kinetic energy} = \tfrac{3}{2}kT$$

where $k = 1.38 \times 10^{-23}\ \mathrm{J K^{-1}}$ is Boltzmann's constant. We can use this equation to get an idea of the typical speed of a water molecule in the steam generated in the boiler of a power station.

Steam is generated in an advanced gas-cooled reactor at a temperature of 645 °C. The average kinetic energy of water molecules at this temperature is given by

$$\text{average } E_k = 1.5 \times 1.38 \times 10^{-23} \times (645 + 273) = 1.9 \times 10^{-20}\ \mathrm{J}$$

The average speed of these molecules is (using $E_k = \tfrac{1}{2}mv^2$; mass of water molecule = 3.0×10^{-26} kg)

$$\text{average speed} = \sqrt{(2 \times 1.9 \times 10^{-20} / 3.0 \times 10^{-26})} = 1125\ \mathrm{m\ s^{-1}}$$

Hence the water molecules are moving at high speeds, and it is their kinetic energy which is extracted as useful mechanical energy by the turbine.

Internal energy and thermal equilibrium

A gas possesses energy, which we can use to do work. The hotter the gas, the greater its energy. This energy is correctly known as its **internal energy** U, and it is made up of the kinetic and potential energies of all its individual particles.

It is easy to picture the molecules of a gas as having kinetic energy, as they are moving around at high speeds. It is perhaps easier to picture the potential energy of particles in a solid. As they vibrate about fixed positions, their energy is continually exchanging between kinetic and potential forms as they speed up, slow down, and reverse their directions of travel.

Thermodynamics deals with macroscopic properties of matter, such as the temperature and internal energy of some gas, but these can always be related back to the underlying microscopic properties of the particles of which the matter consists.

Thermodynamics is also concerned with energy changes, and to understand these we need to develop the idea of thermal equilibrium. We have already discussed the way in which the particles of a gas jostle and collide, and thereby share out their energy. They collide with the walls of their container, and share their energy with it. Put some hot gas in a cold container, and soon the container will itself be hot. (The gas will be cooler.) The gas molecules have shared their energy with the container.

Once the gas and container are at the same temperature, we say they are in **thermal equilibrium** with each other. There is no net transfer of energy between the gas and its container. This is not to say that there is not an exchange of energy between the gas and the container. Particles are still colliding, and sharing energy, but there is a balance (or equilibrium) whereby energy is flowing equally from gas to container and from container to gas.

The idea of thermal equilibrium is essential in thermometry. If you put a cold thermometer under your tongue, you are not in thermal equilibrium with it. The mercury rises; when it stops, thermal equilibrium has been reached, and you can take the correct reading.

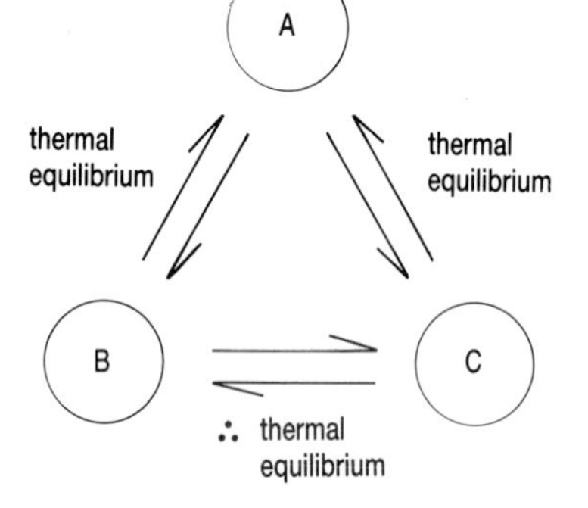

Fig 4.4 A representation of the Zeroth Law of Thermodynamics.

The Zeroth Law and thermal equilibrium

If one body is in thermal equilibrium with another, there is no net flow of energy between them. If a body **A** is in equilibrium with two other bodies **B** and **C** (Fig 4.4), then the Zeroth Law tells us that **B** and **C** are in thermal equilibrium with each other.

This law was an underlying assumption in the early development of the First and Second Laws of Thermodynamics, but it was not clearly stated until the other two laws were well established – hence the name 'Zeroth Law'.

This law is sometimes known as the 'mother, baby and bath-water' law. If a mother wants to be sure that the bath-water is at the right temperature for her baby, she tests its temperature with her elbow, which is very sensitive to temperature. She knows that she is at the same temperature as the baby; if her elbow is in thermal equilibrium with the water, she knows that the baby and the water will also be in thermal equilibrium, and screams of distress can be avoided.

QUESTIONS

4.1 A power station has an output of 600 MW and an overall operating efficiency of 31%. What is the rate at which it uses its energy input? If this is supplied from coal which, when burnt, gives 37 000 kJ per kilogram, at what rate does the power station use coal?

4.2 A hot water tank is used with a 2 kW immersion heater controlled by a thermostat so that the temperature of the water is maintained at 60 °C. On a hot day when the external temperature is 30 °C it is found that the heater is switched on for 20 seconds every quarter of an hour when no hot water is being used.

(a) At what rate does heat energy escape through the lagging jacket?

(b) What would you expect the ON – OFF cycle of the heater to be when the temperature of the surroundings is 0 °C?

4.3 **(a)** What is the average kinetic energy of a molecule in the atmosphere when the air temperature is 17 °C?

(b) The mass of an oxygen molecule is 5.3×10^{-26} kg. What is the speed of an oxygen molecule which has the average kinetic energy calculated in **(a)**?

(c) A nitrogen molecule has 7/8 the mass of an oxygen molecule. What will be the speed of a nitrogen molecule which has the same kinetic energy as the oxygen molecule considered in **(b)**?

4.4 THE FIRST LAW

Usually, in making use of the ideas of thermodynamics, we are concerned with energy changes. If a system is not in thermal equilibrium, energy will flow from one place to another. What can we say about such energy changes?

The First Law of Thermodynamics is in part a statement of the Principle of Conservation of Energy, as applied to thermodynamic systems:

> 'The total energy of a thermodynamic system remains constant, although it may be transformed from one form to another.'

In itself, this statement is of great and fundamental significance, and yet it needs to be adapted to make it useful in particular circumstances. We may want to know what this tells us about, say, the gas expanding in the cylinder of a petrol engine. Let us think about a mass of gas in an enclosed container. How might we increase the temperature of this fixed mass of gas? There are two ways, and these are represented in Fig 4.5.

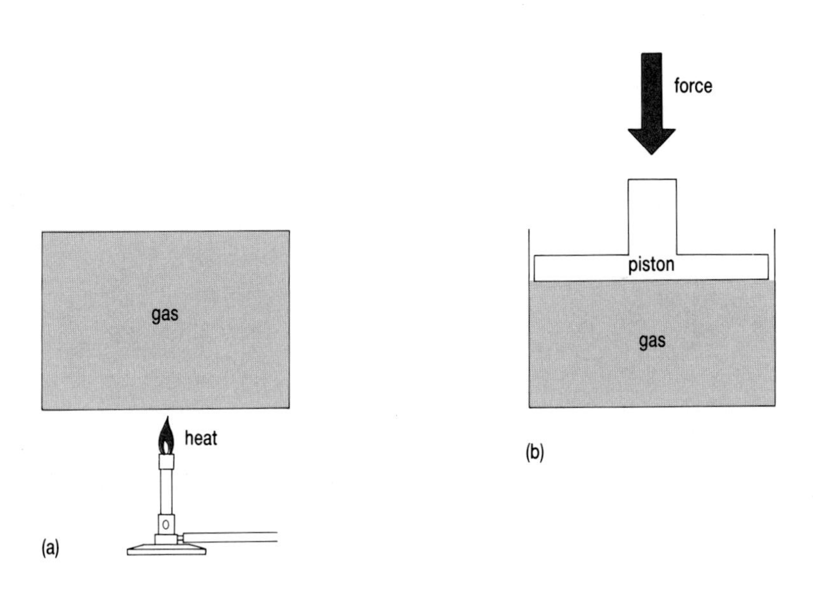

Fig 4.5 Two ways of raising the internal energy of a gas: **(a)** by heating it; and **(b)** by doing work on it.

Firstly, we could heat the gas. If we heat the container, then the gas and container are no longer in thermal equilibrium, and energy will flow from the container to the gas. The energy input as heat we will call ΔQ. Secondly, we could do work on the gas by compressing it. If you have ever pumped up a bicycle tyre, you may have noticed that the tyre and the pump get hot. That is because the air in the tyre becomes hot as you do work on it. Fig 4.5(b) shows this. The piston is forced downwards, and kinetic energy is transferred to the air molecules in just the same way that a moving cricket bat can transfer kinetic energy to a ball with which it comes into contact. The work done on the gas we will call ΔW.

A First Law equation

The energy put into a gas as heat or as mechanical work goes to increase the energy of the particles of the gas; that is, the internal energy U of the gas is increased by an amount ΔU. Energy cannot disappear; it is conserved in a closed system.

Hence we can express the energy changes in an equation:

$$\Delta U = \Delta Q + \Delta W$$

and we can rewrite this equation in words:

gain in internal energy = energy supplied as heat + work done on gas

(You should beware: you may find this equation written in slightly different ways. Sometimes ΔW is used to mean the work done *by* a gas in expanding, in which case it may appear on the other side of the equation, or have a minus sign.)

The First Law says that you can increase the internal energy (and hence the temperature) of a gas by heating it or by doing work on it. The heat supplied and work done do not disappear; they become part of the internal energy of the gas, and could, in principle, be recovered.

Compressing a gas

A gas is compressed at a constant pressure of 5.0×10^5 Pa from a volume of 3.0×10^{-4} m^3 to a volume of 1.0×10^{-4} m^3. During this process its internal energy increases by 70 J. How much heat was lost during the compression?

work done on gas = force × distance moved

= pressure × area × distance moved

= pressure × change in volume

$= 5.0 \times 10^5 \text{ Pa} \times (3.0 - 1.0) \times 10^{-4} \text{ m}^3$

= 100 J

So using the First Law

increase in internal energy	= work done on gas	+ heat supplied to gas
70 J	= 100 J	+ heat supplied to gas
i.e.	Heat supplied to gas = – 30 J	
or	Heat lost from gas = 30 J	

A note about heat

We need to be careful about the way in which we use the word 'heat' as a noun. If a gas is hot, we might loosely say that it has a lot of 'heat' in it. This is not strictly true, since the gas might have become hot by being compressed, rather than by being heated. We could get the energy out of the gas by allowing it to expand and do work, perhaps pushing a piston.

In a diesel engine, the mixture of fuel and air is compressed so much that it becomes hot enough to explode, without being heated. It expands, pushing out the piston and thereby doing work. It cools down without having heat removed. It is more accurate to refer to heat as energy which is moving from somewhere hotter to somewhere colder, that is, as a result of a lack of thermal equilibrium. Heat is the flow of energy as a result of a temperature difference.

QUESTION

4.4 Think about a balloon filled with air. Which of the following processes would lead to a change in the internal energy of the air which initially fills the balloon? In each case, explain your answer.

(a) Putting it in a refrigerator.
(b) Tying it to the back of a fast-moving car.
(c) Carrying it up a hill.
(d) Squeezing it.
(e) Bursting it.
(f) Putting it in a car, which then accelerates.

4.5 THE SECOND LAW

Heat engines, as we have seen, are a general type of engine, used to convert heat energy to mechanical energy. As you found in answer to question 3.4, they are usually rather inefficient, with efficiency below 50%. The Second Law of Thermodynamics explains why this is so.

A car is a particular example of a machine which uses a heat engine as its source of power. Fig 4.6 shows diagrammatically the energy changes measured for a car travelling at 40 miles per hour along a level road. Energy is available from the fuel at a rate of 72 kW; however, only 9 kW of this is used in overcoming the frictional resistance of the air and the road.

This diagram shows two things very clearly. Firstly, most of the energy supplied to the car is dissipated as heat. Even the useful energy, used to overcome friction, results ultimately in heat. Secondly, the heat engine itself is grossly inefficient. It converts only 20% of the energy from the fuel into other useful forms of energy. So what does this tell us about thermodynamics?

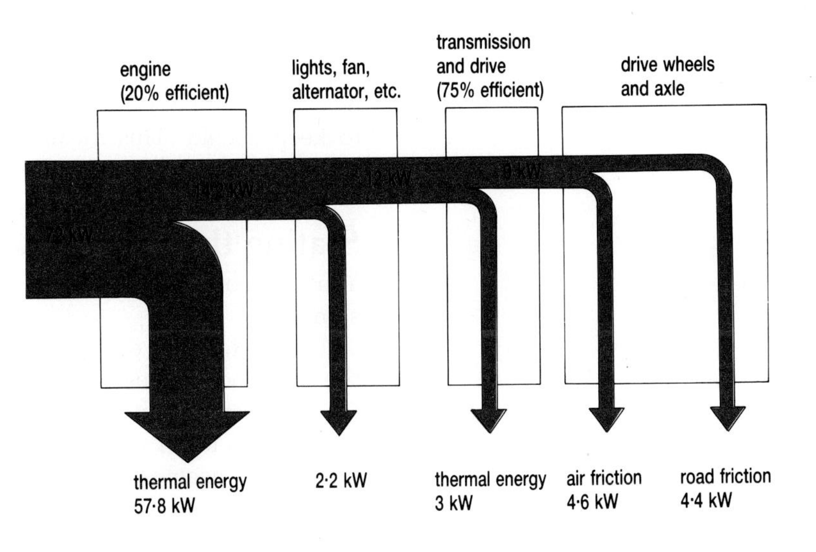

Fig 4.6 The flow of energy from the engine of a car.

The first thing is that heat is a low-grade form of energy. As energy tends to dissipate, it usually ends up as heat. We can make machines to convert energy from one form to another, but poor design leads to wastage of energy as heat. The best electric motor might be 99% efficient; the remaining 1% of energy supplied ends up as heat in the windings.

No doubt you are already familiar with this idea of heat as a degraded form of energy. But the Second Law of Thermodynamics has something much more important to say than this. In effect, it says that it is not possible to make a machine to convert heat energy to mechanical energy with perfect efficiency. A heat engine *cannot* be 100% efficient.

Here, we are not talking simply about bad design. The Second Law is a much more serious limitation than that. It says that there is a theoretical limit to the efficiency of heat engines. In Chapter 5, you will find out how to calculate this limit.

The steam turbine of a power station is a heat engine. Its efficiency is limited to about 60%. In practice, power stations usually have an overall efficiency of about 40%. This may seem like a poor performance; however, there are practical considerations to be taken account of, as well as the Second Law, and these further depress the achievable efficiency. You should note that, at the start of the twentieth century, power stations were only about 5% efficient, and the improvement since then is a triumph of engineering!

Temperature and efficiency

We saw earlier that any heat engine requires a hot reservoir and a cold reservoir. Heat flows from the hotter to the colder and, as it does so, mechanical energy is extracted. From the Second Law we can deduce that the theoretical efficiency of a heat engine can be increased by raising the temperature of the hot reservoir, and lowering the temperature of the cold reservoir.

Consequently, all heat engines work better when they are hot. Engineers work to produce new materials, for example for use in car engines, which will operate at higher temperatures. Ceramic parts are becoming commonplace in car engines, as they remain rigid at temperatures when metal parts start to soften and deform. Power station designers seek new materials for use in boilers, which will not corrode on contact with high temperature, high pressure steam.

Similarly, a heat engine is most efficient when its cold reservoir is kept at a low temperature. For a car engine, this means that designers must ensure that its cooling system, which removes the 80% of energy which becomes heat, is very effective. (The temperature of the cold reservoir cannot be below the temperature of the surroundings, unless extra energy is supplied to keep it cold. This means in practice that the temperature of the cold reservoir cannot be less than about 300 K.)

Stating the Second Law

You may come across the Second Law stated in a variety of ways; here are some.

'Energy tends to dissipate as heat.' This is not a very useful statement of the law, since it does not make clear that energy changes involving heat are fundamentally limited in their efficiency.

'Heat can never pass spontaneously from a colder body to a hotter body.' Of course, heat can be forced to flow from cold to hot; this is what a refrigerator does. Energy must be supplied in order to do this.

'No process is possible which results in the extraction of an amount of heat from a reservoir and its conversion into an equal amount of mechanical work.' Such an impossible process is shown schematically in Fig 4.7. A cold reservoir is needed (the heat has to be flowing to somewhere) and some heat must reach the cold reservoir.

This last statement implies that no heat engine can be 100% efficient.

Why a limit to efficiency?

The Second Law may seem rather dispiriting. We cannot design a heat engine with efficiency much greater than about 50%. Most practical heat engines perform worse than this. How does the law come about?

We have seen that the internal energy of a gas represents the kinetic energy of its particles as they move about randomly. It is this very randomness which is at the root of the problem. In trying to convert heat to mechanical energy, we are trying to convert random motion into organised, directed motion – Fig 4.8. We are trying to convert the random, haphazard motion of gas particles in a cylinder into the linear motion of a piston, and hence into the linear motion of the car itself. And converting randomness into order is difficult. It cannot be done with 100% efficiency.

Our experience of the universe tells us that randomness and disorder tend to increase. Indeed, this is another way in which the Second Law is sometimes stated. Scientists quantify the amount of disorder in a system as its **entropy**. In any thermodynamic process, entropy increases. For example, if you drop an ordered pack of cards on the floor, it ends up in disorder. You know intuitively that dropping a shuffled pack of cards is unlikely to leave it lying in suit order on the floor.

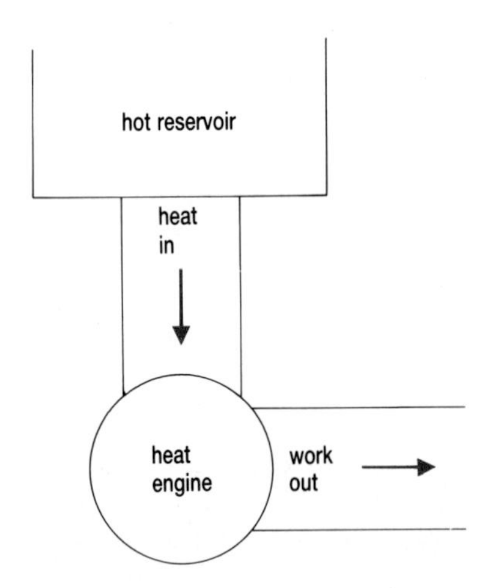

Fig 4.7 An impossible heat engine.

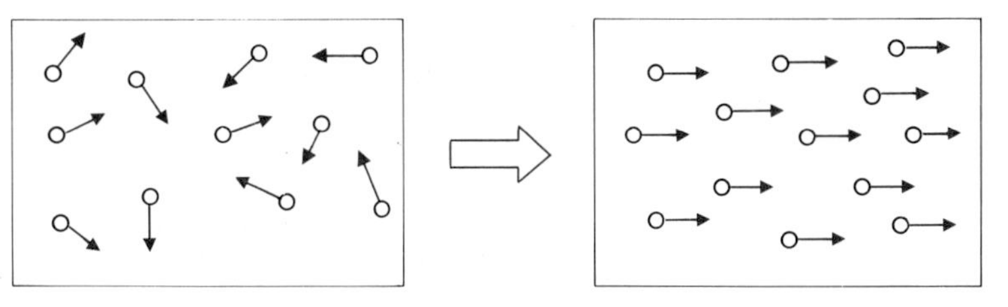

Fig 4.8 A heat engine converts the random motion of the particles of a gas into ordered, linear motion.

The Third Law

There is a Third Law of Thermodynamics, rather different from the other two. This says that the entropy (or disorder) of a substance is zero when the temperature reaches absolute zero.

A consequence of this is that absolute zero is a temperature which cannot be achieved. If we try to reduce the entropy of a substance by cooling it, we can only do so with less than 100% efficiency. We can try again, and cool it further, but some disorder will still remain.

You will find in Chapter 5 that a heat engine can be 100% efficient if its cold reservoir is at absolute zero. Since this temperature cannot be reached, a 100% efficient heat engine is entirely impossible.

Fig 4.9 This Rolls Royce RB 211 – 535EU aero engine gives an output thrust of 178 kN and engines of this type have completed over a million flying hours with an outstanding reliability record.

QUESTIONS

4.5 **(a)** Assuming that the average kinetic energy of a molecule in a liquid such as water is given by $E = \frac{3}{2}kT$ find the internal kinetic energy of a litre of water at 20 °C.
(Mass of a litre of water = 1.00 kg
Mass of a water molecule = 3.0×10^{-26} kg)

(b) The mass of all the water in the oceans of the Earth is approximately 1.5×10^{21} kg. Estimate the internal kinetic energy of all this water.

(c) Why it is impossible to make use of this vast store of energy to heat our homes or drive our transport systems?

4.6 A gas is compressed by doing 6000 J of work on it. At the same time as it is being compressed, 7000 J of heat is extracted from it. What is the resultant change in its internal energy? Does the temperature of the gas rise or fall?

4.7 In an adiabatic change no heat enters or leaves a system. The power stroke of an internal combustion engine can be considered as an adiabatic change and in one particular car engine the internal energy of the gas in a cylinder falls by 800 J during its power stroke. How much work is done during the power stroke?
If the heat capacity of the gas in the cylinder is 1.6 J K^{-1} by how much does the temperature of the gas fall during the power stroke?

4.8 Here are two situations where a knowledge of the laws of thermodynamics can help our understanding:

(a) A class is visiting a watermill. One student suggests that the turning wheel could be used to pump water back up to the millpond, so that there would never be a shortage of water, and the mill could operate continuously.

(b) A group of students are discussing how transport could be made less polluting. One suggestion is to use electric cars, as these do not emit fumes, and their motors can be very efficient. They would be easy to charge from the National Grid.

In each case, explain why the student's suggestion is incorrect.
(i) Give an explanation in everyday terms, such as you might give to a friend who is not studying science. (ii) Give a second explanation, referring to the laws of thermodynamics and any other relevant physical principles.

SUMMARY

The laws of thermodynamics tell us about processes involving energy changes. The Zeroth Law describes what we mean by 'thermal equilibrium'. The First Law is a statement of the Principle of Conservation of Energy.

The Second Law relates to heat engines. It says that there is always inefficiency in a process in which energy from a hot reservoir is converted into mechanical work. This means that there are strict limits to the efficiency which can be achieved by any practical heat engine.

Chapter 5

PRACTICAL HEAT ENGINES

In the previous chapter the basic laws of thermodynamics were discussed. In this chapter the laws will be applied to the gases present in internal combustion engines, enabling the factors affecting the efficiency of the engines to be determined. We shall be looking, therefore, at ways in which a mechanical engineer might attempt to increase engine efficiency.

LEARNING OBJECTIVES

After studying this chapter you should be able to:

1. appreciate why heat engines operate in cycles;
2. use indicator diagrams;
3. calculate the theoretical efficiency of a heat engine;
4. describe the principle of operation of a heat pump and a refrigerator.

5.1 CYCLICAL HEAT ENGINES

Fig 5.1 At take off a rocket burns fuel at an enormous rate. The chemical energy stored in the fuel is converted into increasing kinetic and potential energy of the rocket and a great deal of internal energy.

The concept of a cycle

Providing the power for movement can be achieved in many different ways. Walking and cycling are two ways in which the power is supplied directly by human muscles. Animal power together with wind and water power were the only sources of energy for machines until the beginning of the 18th century when Savery (1698) invented what he called a fire engine, that is, an engine which uses fire in order to pump water. Newcomen (1705) developed this into a pump for extracting water from mines.

When any motion is produced changes occur in the state of the device producing the motion. For example, if a compressed spring is used to shoot a ball on a pin table the spring expands and loses potential energy as it supplies kinetic energy to the ball. If water in a hydroelectric reservoir falls and drives a turbine there is less water left in the reservoir after making the turbine move.

The total useful energy output of a car in its lifetime could be that supplied by a power of 50 kW for 5000 hours: a total energy of around 10^{12} J. In no way can this quantity of energy be stored in a newly purchased car so the practical alternative is to work in a **cycle** of operations. At one stage, energy is supplied by the device to produce motion and at a later stage some process is used to return to the starting position so that more energy can be supplied to produce motion.

One example of cyclical energy conversion would be an electrically operated car. The charged battery is used to drive an electric motor and after the battery is discharged it must be recharged before it can be used again. In practice, heat engines work in a cycle of operations with the result that the limitations of the laws of thermodynamics apply and some of the heat energy input is converted into work and some of it is wasted.

Indicator diagrams

The cyclical nature of heat engines can easily be seen on a graph of pressure against volume if the engine is one in which a gas is the working substance. These diagrams are called **indicator diagrams**. The volume is always plotted on the x-axis and the pressure on the y-axis as shown in Fig 5.2. The direction of any change in volume which occurs needs to be shown. Fig 5.2 shows a gas expanding at constant pressure and doing work on some external system as a result. Here the gas is at a pressure of 1.0×10^6 Pa and the volume changes from 0.0005 m^3 to 0.0025 m^3. The area shaded gives the work done by the gas on its surroundings, so here the gas does work equal to $p\Delta V$.

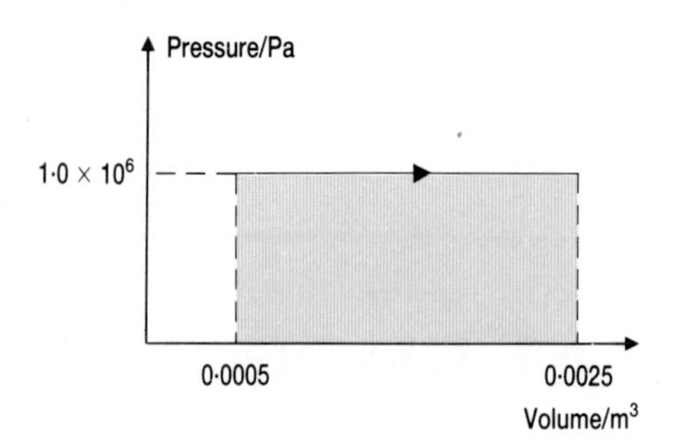

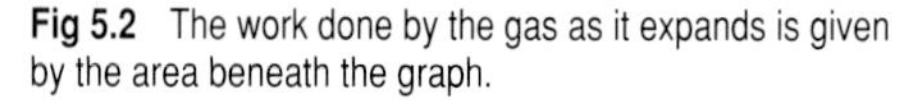

Fig 5.2 The work done by the gas as it expands is given by the area beneath the graph.

$$\begin{aligned}\text{Work done} &= 1.0 \times 10^6 \text{ Pa} \times (0.0025\text{–}0.0005)\text{ m}^3 \\ &= 1.0 \times 10^6 \times 0.0020 \text{ J} \\ &= 2000 \text{ J}\end{aligned}$$

An indicator diagram can be used to find the idealised efficiency of an engine. This is shown by considering the complete cycle of operations given in Fig 5.3. The diagram is idealised but there is an engine, called the Stirling engine, which has a mode of operation similar to this. Air of initial volume 0.0005 m^3 is contained within a cylinder at a pressure of 1.0×10^5 Pa and a temperature of 300 K. (Absolute temperature is always used.) This is represented by point **A**.

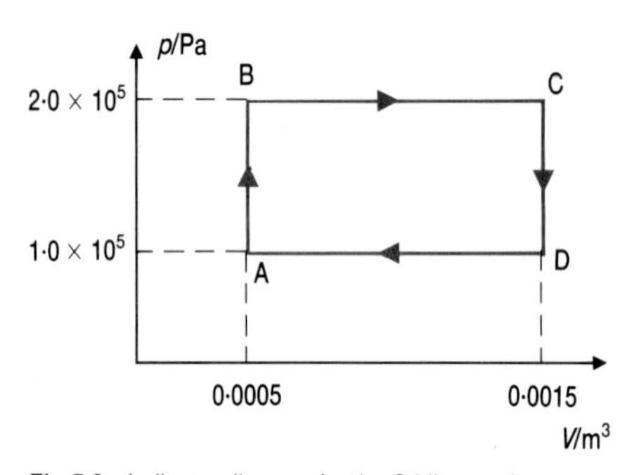

Fig 5.3 Indicator diagram for the Stirling engine.

From the universal gas equation

$$pV = nRT$$

We get

$$1.0 \times 10^5 \times 0.0005 = nR \times 300$$

Since $R = 8.31$ J kg^{-1} mol^{-1} this gives

$$n = \frac{1.0 \times 10^5 \times 0.0005}{8.31 \times 300}$$

$$= 0.020 \text{ mol}$$

Using the universal gas equation for an ideal gas enables the temperature of the gas to be found at points **B**, **C** and **D**. These are quoted in Table 5.1

Table 5.1

	Pressure/Pa	Volume/m^3	Temperature/K
A	1.0×10^5	0.000 50	300
B	2.0×10^5	0.000 50	600
C	2.0×10^5	0.001 50	1800
D	1.0×10^5	0.001 50	900

Next the heat supplied for each stage can be calculated if the specific heat capacities at constant pressure ($C_{p,\,m}$) and constant volume ($C_{v,\,m}$) are used. These are

$$C_{p,\,m} = 29.1 \text{ J mol}^{-1}\text{ K}^{-1} \text{ and } C_{v,\,m} = 20.8 \text{ J mol}^{-1}\text{ K}^{-1}$$

Working to the nearest 5 J gives the values quoted in Table 5.2. In our perfect engine **A** to **B** is taking place at constant volume so the heat supplied is given by

$$\begin{aligned}\Delta Q &= n\, C_{v,\,m}\, \Delta T \\ &= 0.020 \times 20.8 \times 300 \\ &= 125 \text{ J}\end{aligned}$$

B to **C** will also be at constant pressure, so

$$\begin{aligned}\Delta Q &= n\, C_{p,\,m}\, \Delta T \\ &= 0.020 \times 29.1 \times 1200 \\ &= 700\ \text{J}\end{aligned}$$

and similarly for **C** to **D** and **D** to **A**, where the temperatures fall.

Table 5.2

	Heat supplied to gas /J	Work done on gas /J	Increase in internal energy /J
A → B	125	0	125
B → C	700	–200	500
C → D	–375	0	–375
D → A	–350	+100	–250

We saw in Chapter 4 that

work done on a gas = pressure × change in volume

and so we can calculate the work done on the gas at each stage in the cycle (Table 5.2). By applying the First Law of Thermodynamics we can then work out the change in internal energy at each stage.

Efficiency

The indicator diagram given in Fig 5.3 can be used together with Table 5.2 to find the efficiency of this idealised engine by using each part of the cycle.

A→B 125 J of heat supplied to the gas to raise its temperature at constant volume.

B→C a further 700 J of heat supplied while the gas expands at constant pressure and does 200 J of work on its surroundings.

C→D 375 J of heat extracted from the gas to cool it at constant volume.

D→A to return the gas to its starting point 100 J of work have to be done on it and 250 J of heat have to be extracted from it so that its volume falls at constant pressure.

Overall, 825 J are supplied as heat, 725 J are extracted as heat and a net 100 J of work are done by the gas (200 J done by the gas – 100 J done on the gas to compress it).

Having completed the cycle **A** – **D** the gas is ready to start another cycle. Even in this ideal system the 725 J of heat extracted is lost to the system. It cannot be used on the next cycle as part of the 825 J input because it is heat extracted at a low temperature and the 825 J of heat needed for the next cycle has to flow from a high temperature. This is the crux of the problem. Of the 825 J of heat supplied, 100 J only is used to do work and the remainder is wasted. Fig 5.4 is a Sankey diagram which illustrates this.

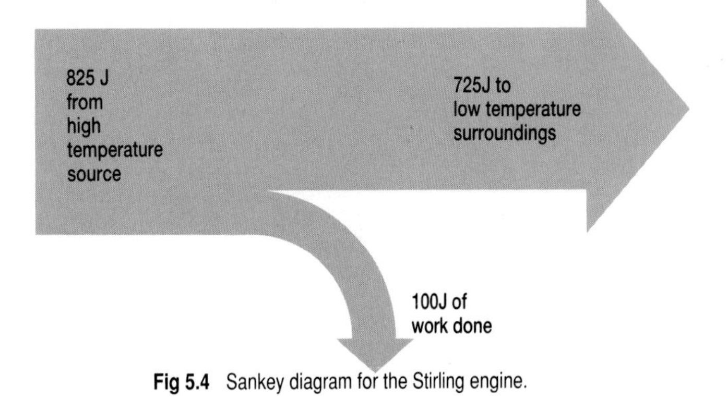

Fig 5.4 Sankey diagram for the Stirling engine.

The efficiency of such a system is defined by the equation

$$\text{efficiency } (\varepsilon) = \frac{\text{net work output}}{\text{total heat input}}$$

and in this case $\varepsilon = \dfrac{100 \text{ J}}{825 \text{ J}}$

$$= 0.12$$

$$= 12\%$$

Note that the efficiency is also given by

$$\text{efficiency} = \frac{\text{heat input} - \text{heat output}}{\text{heat input}}$$

Factors limiting practical efficiency

An engine such as the one just described obviously has a low efficiency and to increase its efficiency would be the aim of any mechanical engineer. Increasing the efficiency of a heat engine can be done in two fundamentally different ways.

1. The theoretical, ideal efficiency can be increased by designing the engine so that it has a higher ideal efficiency. This involves considering the theory of the cycle being used and the temperatures and pressures being used.
2. The practical efficiency can be increased by designing the engine so that friction is low, combustion of fuel is complete, valves are airtight and there is no unnecessary movement of engine parts and gases. It is quite typical for a car engine to have a maximum design efficiency of 60% but to have an actual efficiency of only 30%. It really is difficult to get work from heat.

To deduce the maximum possible efficiency for an engine working between a high temperature T_H and a low temperature T_L is quite difficult but a cycle which does have the maximum possible efficiency is shown in Fig 5.5, and the data for this cycle are given in Tables 5.3 and 5.4. The

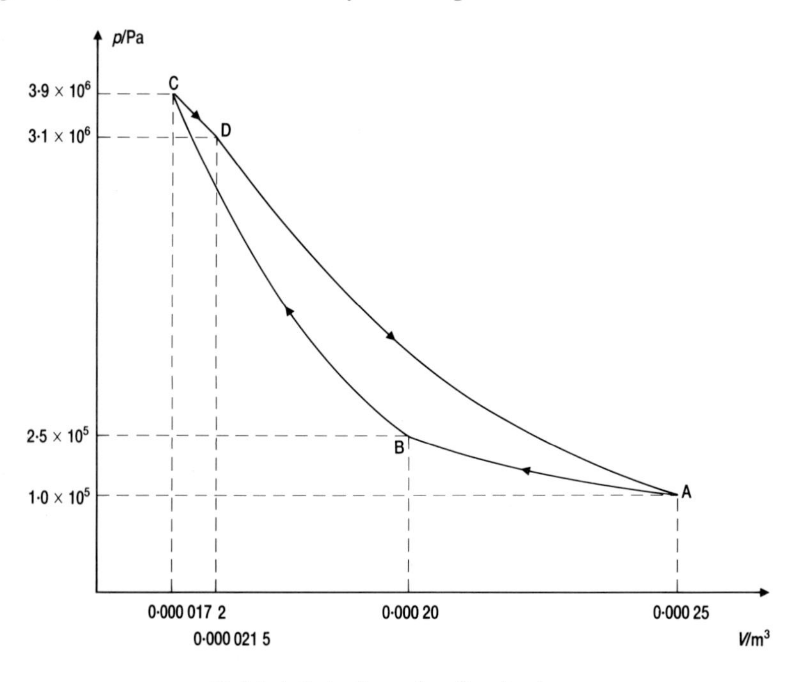

Fig 5.5 Indicator diagram for a Carnot cycle.

engine is assumed to be using 0.0100 mol of an ideal gas and to be working between temperatures of 800 K and 300 K.

Table 5.3

	p/Pa	V/m³	T/K
A	1.0×10^5	0.000 25	300
B	2.5×10^5	0.000 20	300
C	3.9×10^6	0.000 017 2	800
D	3.1×10^6	0.000 021 5	800

Table 5.4

	Heat supplied to gas /J	**Work done on gas /J**	**Increase in internal energy /J**
A→B	–5.6	5.6	0
B→C	0	104	104
C→D	14.9	–14.9	0
D→A	0	–104	–104

The cycle is called a **Carnot cycle** and for an ideal gas consists of:

A→B a compression at constant temperature. During this compression work is done on the gas, but since there can be no change of internal energy at a constant temperature it is necessary for some heat to be extracted.

B→C a further compression during which no heat is supplied or extracted. This is called an **adiabatic compression**.

C→D an expansion at constant temperature.

D→A an adiabatic expansion.

The efficiency of this cycle is found from Table 5.4 using

$$\text{efficiency} = \frac{\text{net work out}}{\text{total heat input}}$$

$$= \frac{(14.9 + 104 - 5.6 - 104)\ \text{J}}{14.9\ \text{J}}$$

$$= \frac{9.3\ \text{J}}{14.9\ \text{J}} = 0.63$$

Another equation for maximum efficiency of a Carnot cycle is

$$\text{maximum efficiency} = 1 - \frac{T_L}{T_H}$$

$$= 1 - \frac{300}{800}$$

= 0.63, which agrees to 2 significant figures.

The equation for the maximum possible efficiency of a heat engine working between a low temperature T_L and a high temperature T_H has extremely important consequences in the field of energy supply. The implications of the equation are often ignored in social and political discussions but the laws of thermodynamics cannot be broken and you should now be able to answer the following questions and also be able to see why the equation is so important.

QUESTIONS

5.1 **(a)** What is the maximum possible efficiency of a steam engine working with steam at a high temperature of 127 °C (400 K) and a low temperature, at which it condenses, of 77 °C (350 K)?

(b) How do the temperatures used need to be changed in order to increase the efficiency? What is the problem with each change proposed?

(c) What would the low temperature need to be to have a 100% efficient engine?

5.2 **(a)** A magnox nuclear power station has cooling water at a temperature of 17 °C and the temperature of the reactor cannot be allowed to go above a temperature of 370 °C or the cans containing the nuclear fuel would melt. What is the maximum theoretical efficiency of the reactor?

(b) The advanced gas cooled reactors use steel cans to hold the nuclear fuel which is in ceramic form. They can use a maximum temperature of 625 °C. What is the maximum theoretical efficiency of such a reactor?

5.3 A power station has a practical efficiency of 33% and a maximum theoretical efficiency of 68%. Suggest reasons why these figures are so different.

Fig 5.6 Ferrybridge C, a 2000 MW coal fired power station.

5.4 A conventional power station (coal fired) has an electrical output of 1200 MW. Its actual efficiency is 36%.

(a) How much heat is provided by the burning coal each second?

(b) How much heat is wasted by the power station each second?

(c) State two ways in which this wasted heat can be removed from the power station.

5.5 The following figures were given by a central heating consultant to a customer who required a central heating system capable of supplying 25 kW.

	Supplied cost /kW h	Efficiency	Cost of heat used /kW h	Cost /h
Gas	1.5p	77%		
Night storage electricity	3p	100%		
Normal price electricity	7p	100%		
Oil	2.4p	68%		

(a) Suggest the meaning of efficiency in this case.
(b) Find the cost of each kW h of heat used and the cost of running each system for an hour.

Fuel consumption

Here again calculations tend to be those which give idealised values. In practice, more fuel will be required than an idealised calculation suggests. Nevertheless, it is useful to know the minimum amount of fuel needed. If nothing else it gives a target for a mechanical engineer to aim at when designing an engine.

If the Carnot cycle referred to earlier is required to give an output power of 400 W then it must undergo

$$\frac{400\text{ W}}{9.3\text{ J}} = 43 \text{ cycles per second}$$

The supply of heat to the engine is 14.9 J per cycle so the rate of heat input is

$$14.9\text{ J} \times 43 \text{ cycles per second} = 640\text{ J s}^{-1} = 640\text{ W}$$

Since many petroleum products give about 45 000 J for each gram burnt this engine would, at minimum rate of consumption, only need

$$\frac{640\text{ J s}^{-1}}{45\,000\text{ J g}^{-1}} = 0.014\text{ g s}^{-1}$$

This is about 70 cm^3 of fuel per hour.

QUESTION

5.6 A car travelling with a speed of 120 kph needs work to be done on it at a rate of 40 kW. Its practical efficiency is 35%. The petrol it uses supplies 3.8×10^7 joules of heat per litre of fuel burnt.
Find:
(a) the rate at which fuel must be supplied to the engine.
(b) the number of litres required for a journey of 100 km. (This is how fuel economy is measured on the continent.)
Discuss what happens to the 40 kJ of mechanical work done on the car each second in terms of the Law of Conservation of Energy.

5.2 HEAT PUMPS

Refrigerators

In a refrigerator like the one shown in Fig 5.7, a gas is compressed in a compressor until it is liquefied. During this process the gas warms up so it is passed through a long, black painted, cooling tube placed outside on the back of the refrigerator. The liquefied gas loses heat by convection and by radiation and its temperature falls to near room temperature. It is then allowed to expand through an expansion valve placed near the icebox of the refrigerator and in doing so it vaporises and its temperature drops, so cooling the icebox. The gas used for refrigerators has for many years been a chlorofluorocarbon (CFC). However, because these gases damage the ozone layer in the atmosphere other gases are now being used. There is, however, still a large quantity of CFC gases in refrigerators around the world and if nothing is done to prevent the escape of the gas it will enter the atmosphere when these refrigerators are scrapped. The problem is what to do with the CFC in fridges which are being scrapped.

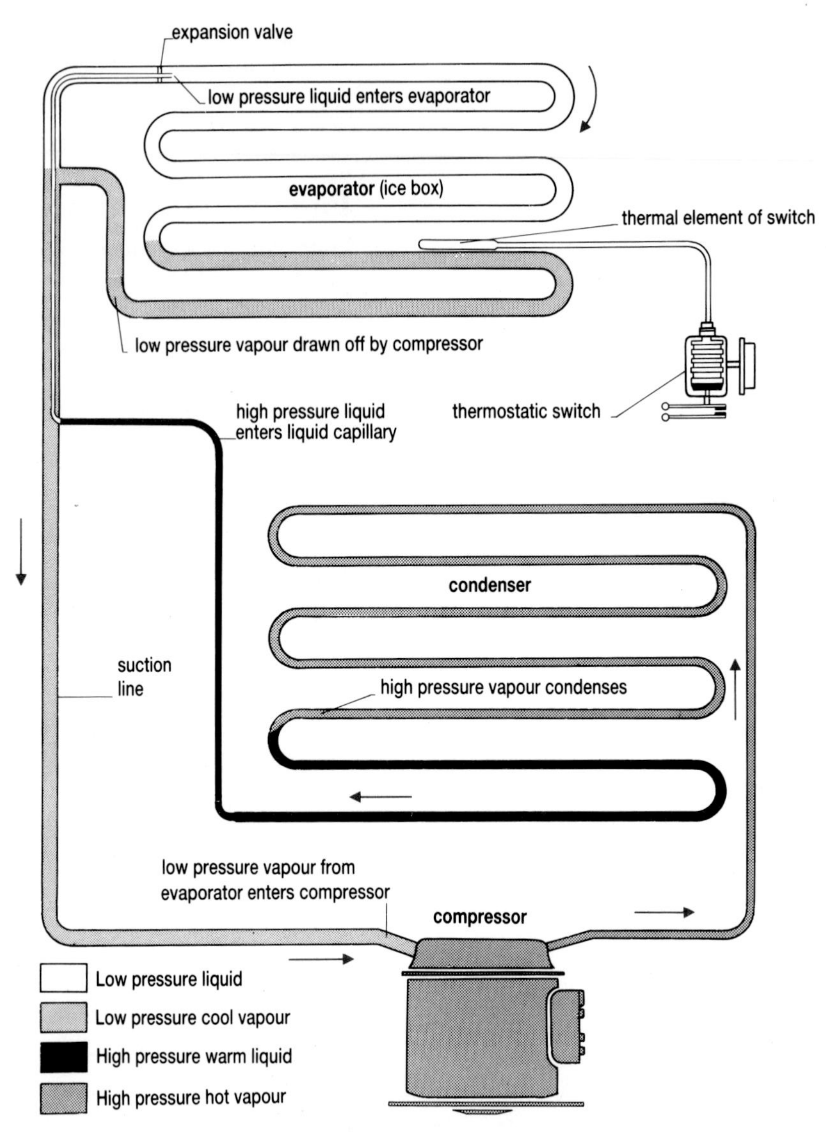

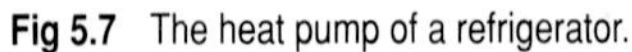

Fig 5.7 The heat pump of a refrigerator.

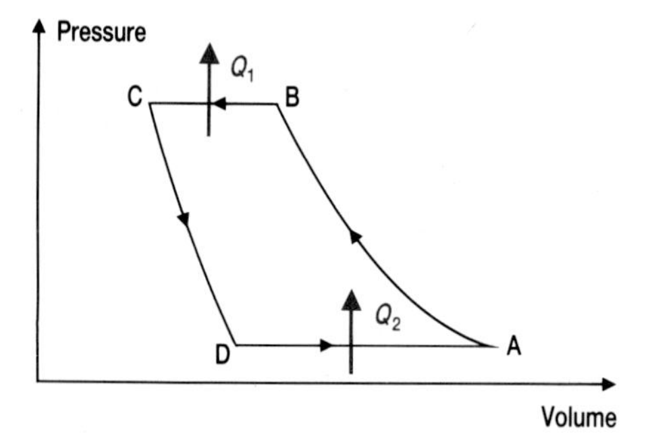

Fig 5.8 Indicator diagram for a refrigerator.

In principle a refrigerator is a heat engine cycle worked backwards. This is shown in Fig 5.8 where the cycle consists of:

A→B an adiabatic compression.

B→C a cooling of the gas at constant pressure: heat is lost to the surroundings.

C→D an adiabatic expansion.

D→A the gas heats up at constant pressure: heat is absorbed from the surroundings.

By making the change **B→C** take place at the back of a refrigerator, see Fig 5.7, heat is lost to the outside of the refrigerator. The coils on refrigerators are quite hot when it is in operation. Change **D→A** takes place in the icebox and heat is absorbed by the gas from the ice box.

The work done on the gas during one cycle is equal to the area enclosed and is shown in the Sankey diagram, Fig 5.9.

$$Q_1 = Q_2 + W$$

where Q_2 is the heat extracted from the food; W is the work done on the gas and Q_1 is the heat delivered to the outside.

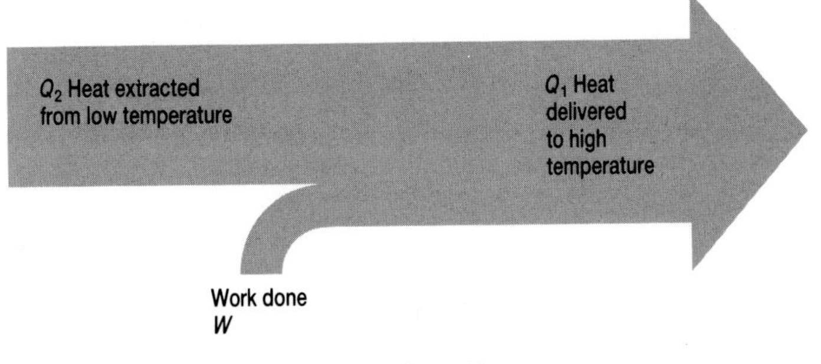

Fig 5.9 Sankey diagram for a refrigerator.

Coefficient of performance

Just as with the heat engine, the Carnot cycle gives the most efficient refrigerator possible. That is, it extracts the maximum amount of heat from the low temperature system for a given amount of work done.

A measure of how good a refrigerator is, is called its **coefficient of performance**, and is given by

$$\text{coefficient of performance} = \frac{\text{heat extracted}}{\text{work done}}$$

$$= \frac{Q_2}{Q_1 - Q_2}$$

Since for a Carnot cycle

$$\frac{Q_1}{Q_2} = \frac{T_1}{T_2}$$

$$\text{coefficient of performance of refrigerator} = \frac{T_2}{T_1 - T_2}$$

where T_1 is the high temperature and T_2 is the low temperature

QUESTION

5.7 (a) A refrigerator is working in a kitchen in which the temperature is 21 °C. The icebox is at a temperature of 0 °C. What is the maximum possible rate of extraction of heat from food in the icebox if the motor is working at a rate of 100 W?

(b) What happens to the numerical size of the coefficient of performance as T_2 falls? At what temperature would the coefficient of performance become zero? What does this answer imply about the practical difficulty of reaching low temperatures?

Heat pumps for heating

The refrigerator pumps heat from a low temperature source to high temperature surroundings using a motor to do the necessary work. This principle is made use of in a heat pump which can be substituted for a central heating system. Using Fig 5.9 again shows that if heat Q_2 is extracted from the low temperature source and work W is done to pump this heat, then heat Q_1 is supplied to the surroundings. The coefficient of performance for a heat pump is slightly different from that for a refrigerator, being given by

$$\text{coefficient of performance} = \frac{\text{heat supplied to high temperature surroundings}}{\text{work done}}$$

$$= \frac{Q_1}{W}$$

$$= \frac{Q_1}{Q_1 - Q_2}$$

What makes heat pumps interesting is that Q_1 can be larger than W. The maximum possible value for coefficient of performance, with a Carnot cycle, is found using

$$\frac{Q_1}{Q_2} = \frac{T_1}{T_2}$$

and is

$$\text{coefficient of performance} = \frac{T_1}{T_1 - T_2}$$

A heat pump uses the air outside a building as its low temperature source. The temperature of the external air is 0 °C. Heat is pumped into the building at a rate of 6000 W using a 2000 W motor and the temperature inside the building is 20 °C.

How does the coefficient of performance of a heat pump compare with the maximum possible coefficient of performance?

$$\text{Practical coefficient of performance} = \frac{6000\text{ W}}{2000\text{ W}} = 3$$

Maximum possible coefficient of performance

$$= \frac{T_1}{T_1 - T_2} = \frac{(20 + 273)}{(20 + 273) - 273} = \frac{293}{20}$$

$$= 14.7$$

These figures are very different because

- a Carnot cycle is not feasible in practice;
- an ideal cycle is not attained in practice: there is a great deal of irreversibility in any cycle and irreversibility always causes a reduced efficiency;
- heat exchange between the outside air and the heat pump is poor;
- some heat is lost.

The example shows that, theoretically, a great deal of improvement can still be made to heat pumps. Even so, it is much more efficient to use a 2000 W motor to provide 6000 W of heat than to use 2000 W of electrical energy to produce 2000 W of heat from an electric fire. Practical heat pumps are available commercially, Fig 5.10, and coefficients of performance of up to 5 are possible, although good thermal contact between the outside air, which cools down, and the input of the heat pump is important. Unfortunately, the lower the external temperature the lower the coefficient of performance becomes, so on a very cold day the heat pump is less able to supply the necessary heat than it is on a hot day – when it is probably not required. In some cases the input heat exchanger is immersed in a stream or river. This enables the heat to be extracted much more readily.

Fig 5.10 This heat pump is a Lennox HP 17. It is able to give a net output of 22 kW with a power supply requirement of only 5.3 kW.

5.3 OTHER PRACTICAL HEAT ENGINES

4-stroke petrol engines – Otto cycle

Although some larger engines have more than four cylinders, the most common type of car engine has four cylinders, as shown in Fig 5.11.

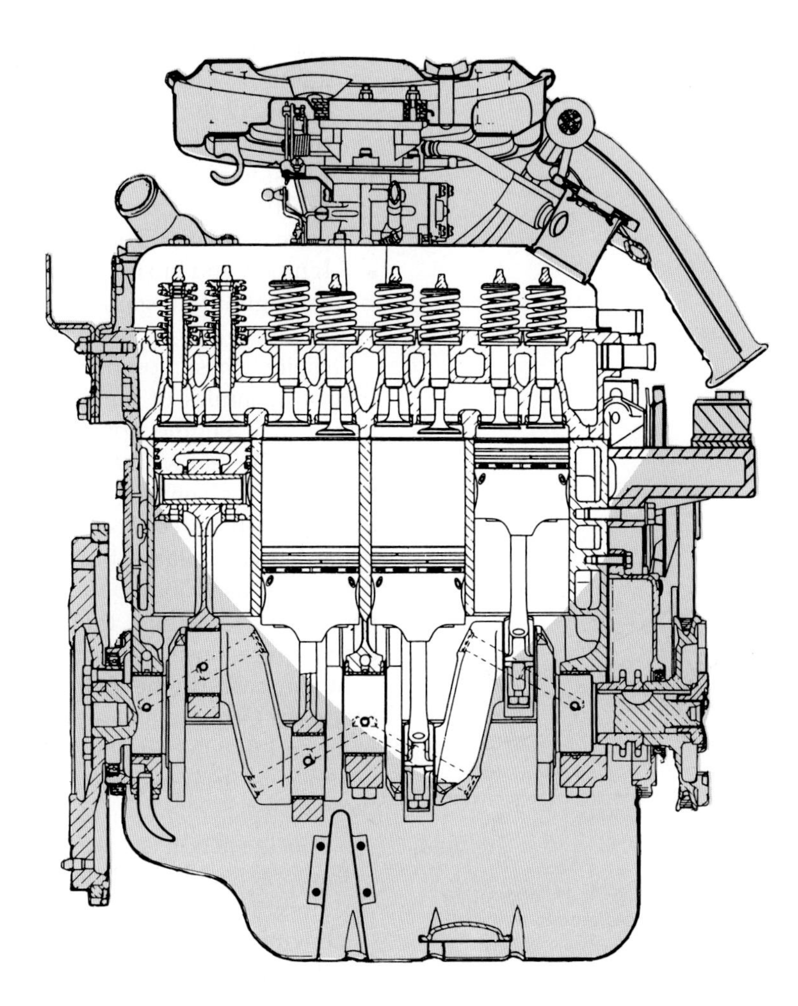

Fig 5.11 A 4-stroke petrol engine having 4 cylinders so that each piston is undergoing a different stroke at any instant. In this diagram the two valves are clearly visible above each cylinder. The springs keep the valves shut unless they are forced open.

Each cylinder is performing a different stroke from any other cylinder at any moment. The normal stroke sequence is shown in Table 5.5

Table 5.5

Piston in Cylinder 1	Piston in Cylinder 2	Piston in Cylinder 3	Piston in Cylinder 4
induction	compression	exhaust	power
compression	power	induction	exhaust
power	exhaust	compression	induction
exhaust	induction	power	compression

This sequence has the advantage that there is always one piston on its power stroke. Also there is least amount of vibration produced within the engine since the two outer pistons (1 and 4) go up and down in phase with one another and the two inner pistons (2 and 3) are also in phase with one another and in antiphase to pistons 1 and 4. This double see-saw arrangement means that there is no tendency for the front or the back of the engine to rise or fall at any time.

The thermodynamics of a 4-stroke cycle in a car engine are complicated by several factors. Some of these are

- The fuel burns in air during the cycle within the cylinder. The number of moles of gas present is not therefore constant.
- The cycle takes place very quickly so the gases present before and after ignition swirl through the cylinder in a complex manner. Kinetic energy of the gases present is not normally considered thermodynamically.
- Considerable temperature gradients are present during a cycle, so it is practically impossible to deal with the temperature of the gas as if it were constant throughout the gas.
- Ignition and burning take a finite time, so pressures will vary throughout the gas at any moment.

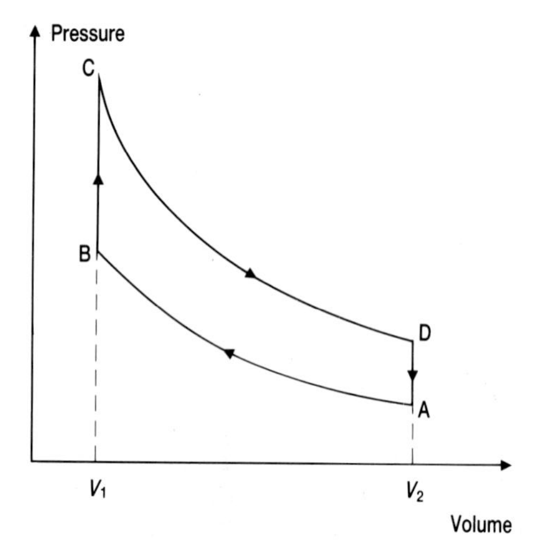

Fig 5.12 Indicator diagram for the Air Standard Otto cycle. This is an idealised diagram which approximates to the diagram for a petrol engine.

Despite these problems it is still enlightening to examine the thermodynamics of an idealised engine cycle and then to consider how near a real engine approaches the ideal. This is a common method in science. The real world *is* complex. Ideal situations are first considered and the answers given can then be refined by a series of steps as extra practical details are included. For example, a simple pulley system can be dealt with using Newton's laws, but as greater accuracy is required first the mass of the pulley itself may be included, then its kinetic energy of rotation, then friction at its bearings, then the fact that the string passing over it has mass, then air resistance, etc., etc.

In the case of the petrol engine the idealised cycle is called the **Air Standard Otto** cycle. It assumes that the working substance is a fixed mass of air and that the air obeys the gas laws throughout the cycle.

The cycle itself is shown in Fig 5.12 and consists of the following stages:

1. The induction stroke. This takes place at **A**. The inlet valve of the engine is open and the piston falls. The fixed mass of air, theoretically always at atmospheric pressure, is simply transferred from outside the engine to inside the space in the cylinder. Note that this assumes that the piston falls very slowly so that the pressure inside and out is the same. In practice it is not possible to wait for this to occur so the pressure inside the cylinder is lowered to increase the speed with which the gas enters the cylinder.
2. The compression stroke, **A→B**. This is an adiabatic compression with both valves closed. No heat should enter or leave the cylinder during this stroke.
3. Ignition, **B→C**. An instantaneous supply of heat is given to the air – in practice by igniting the petrol/air mixture with a spark from the spark plug.
4. The power stroke, **C→D**. An adiabatic expansion of the air. Work is done on the piston by the expanding gas.
5. Cooling, **D→A**. The gas is cooled instantaneously at constant volume to bring it back to its original temperature.
6. Exhaust stroke. At **A**, the exhaust valve is opened and the exhaust gases are transferred to outside the engine at constant volume and pressure.

In practice stages 5 and 6 are combined together. What is required after the power stroke is to get back to **A** as economically as possible. Steps 5 and 6 do this most efficiently. Any other system involves more inefficiency.

An outline theory of this cycle is as follows:

$$\text{heat absorbed } \mathbf{B \to C} = Q_{\text{in}} = n\, C_{\text{v, m}}\,(T_{\text{C}} - T_{B})$$

$$\text{heat rejected } \mathbf{D \to A} = Q_{\text{out}} = n\, C_{\text{v, m}}\,(T_{\text{D}} - T_{\text{A}})$$

$$\text{efficiency } \varepsilon = \frac{\text{work done}}{\text{heat absorbed}} = \frac{Q_{in} - Q_{out}}{Q_{in}}$$

$$= \frac{n\, C_{v,m}\,(T_C - T_B) - n\, C_{v,m}\,(T_D - T_A)}{n\, C_{v,m}\,(T_C - T_B)} = 1 - \frac{T_D - T_A}{T_C - T_B}$$

Further analysis shows that the ratio

$$\frac{T_D - T_A}{T_C - T_B}$$

is related to the compression ratio of the engine, r. This is the ratio of

$$\frac{\text{maximum volume within cylinder}}{\text{minimum volume within cylinder}} = \frac{V_2}{V_1}$$

The relationship is

$$\frac{T_D - T_A}{T_C - T_B} = r^{-0.4}$$

and hence the efficiency is given by

$$\varepsilon = 1 - r^{-0.4}$$

If the above equation is used to find the efficiency of a car of compression ratio 8.5, we get

$$\text{efficiency} = 1 - r^{-0.4} = 1 - 8.5^{-0.4}$$

$$= 1 - 0.425 = 0.575 \text{ or } 57.5\%$$

Fig 5.13 The Rolls Royce K60 Diesel engine.

The theoretical efficiency is increased if r is increased. For example, for r = 20, ε = 70%. Only if r is infinite is the efficiency theoretically 1, but high compression ratios mean that when the compression stroke occurs the temperature of the petrol/air mixture rises above the ignition point of the petrol and the mixture pre-ignites. This causes tremendous stresses internally in the engine, as the piston is being pushed down on one of its up strokes. This could, in theory, make the engine, and the car, go backwards suddenly. Since this is not possible the piston does continue to rise against the exploding petrol and a knocking sound results. Continued knocking can cause considerable engine damage.

Diesel engine

One way to overcome the problem of pre-ignition is not to have fuel in the cylinder during the compression stroke. The diesel engine is the most common form of such an engine. Its indicator diagram, Fig 5.14, shows the following:

1. The induction stroke: basically as for an Otto cycle. Ideally it takes place at constant pressure, volume and temperature. Air is transferred from outside to inside the cylinder through the inlet valve.
2. The compression stroke, **A→B**. This is an adiabatic change to a very small volume. The compression ratio is typically about 20. As a result of this high compression the temperature of the air rises to a high value.
3. Fuel injection, **B→C**. When diesel oil is injected in atomised form into the high temperature air in the cylinder it burns without needing a spark to ignite it, and pushes the piston down. Ideally, the fuel is injected at a rate that keeps the pressure constant as the piston moves.
4. **C→D**: Power stroke. Once all the fuel is burnt the piston will continue

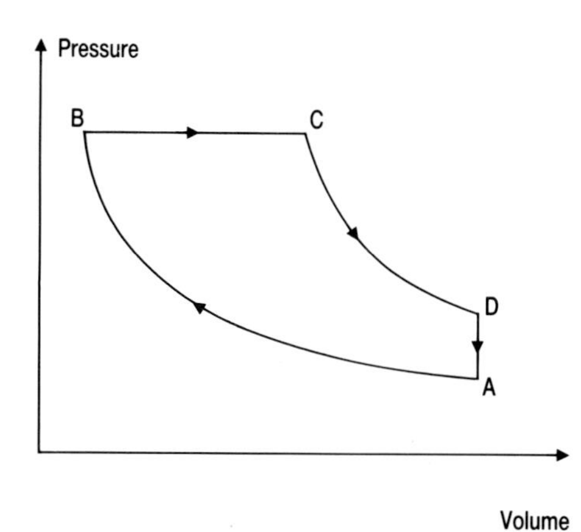

Fig 5.14 The indicator diagram for a diesel engine.

to move down whilst an adiabatic expansion takes place.

5. **D→A**: Cooling and exhaust. As with the Otto cycle, this is a way of rejecting heat to the exhaust and getting back to the starting point so another cycle can occur.

ASSIGNMENT

Data analysis

The following data, modified to some extent for simplicity, refer to a Rolls Royce Motor Cars Ltd. Diesel Engine Type K60. This is an engine much used in military applications.

Number of cylinders	6
Maximum volume within each cylinder	1.10×10^{-3} m^3
Compression ratio	16 to 1
Mass of engine	757 kg
Mass-to-power ratio	4.85 kg kW^{-1}
Power-to-volume ratio	225 kW m^{-3}
Power rating	165 kW at 3750 r.p.m.
Maximum torque	488 Nm at 2500 r.p.m.

Table 5.6

	p/Pa	V/m^3	T/K
A	1.00×10^5	11.0×10^{-4}	300
B	48.5×10^5	0.69×10^{-4}	910
C	48.5×10^5	1.51×10^{-4}	2000
D	3.01×10^5	11.0×10^{-4}	904

Table 5.7

	Work done on gas /J	Heat supplied to gas /J	Increase in internal energy /J
A→B	560	0	560
B→C	–400	1400	1000
C→D	–1010	0	–1010
D→A	0	–550	–550

The cycle is the one in Fig 5.14, and pressure, volume and temperature values for this engine at A, B, C and D are shown in Table 5.6. The application of the First Law of Thermodynamics to each part of the cycle gives the details shown in Table 5.7.

1. The universal gas constant R is 8.31 J mol^{-1} K^{-1}. Use this fact to find how many moles of gas are present in one cylinder during a cycle. (The introduction of fuel is assumed to make a negligible difference to the figure so the value will be constant throughout the cycle.)
2. Why are the mass to power ratio and the power-to-volume ratio important for a designer who wishes to incorporate an engine such as this into a vehicle?
3. Use the graphs of torque and power against speed, Fig 5.15(a) and (b), to explain why the maximum torque does not occur at maximum speed.
4. How is the value 0.69×10^{-4} m^3 obtained for the volume of gas at B?
5. The molar heat capacity at constant volume for a gas such as air is 20.8 J mol^{-1} K^{-1}. Show that heating the gas in a cylinder in this engine requires 0.918 J K^{-1}. Hence show that the increase in the internal energy in moving from D to A is – 550 J.
6. Explain why the increase in internal energy in moving from A to B is 560 J. (There is no need to calculate the work done by the gas during this compression.)
7. Calculate the net work done during the one cycle shown on the indicator diagram, the heat energy supplied during the cycle and hence the theoretical efficiency.
8. Calculate the net work actually done per cycle when the engine is giving 1565 kW output. The engine has 240 injections of fuel per second at this speed.
9. Use Fig 5.15(c). What is the fuel consumption at this speed? How much fuel is injected into the cylinder on each occasion?
10. How much heat does this amount of fuel supply?
11. What is the calorific value of the fuel? That is, how much energy is supplied per kilogram of fuel?

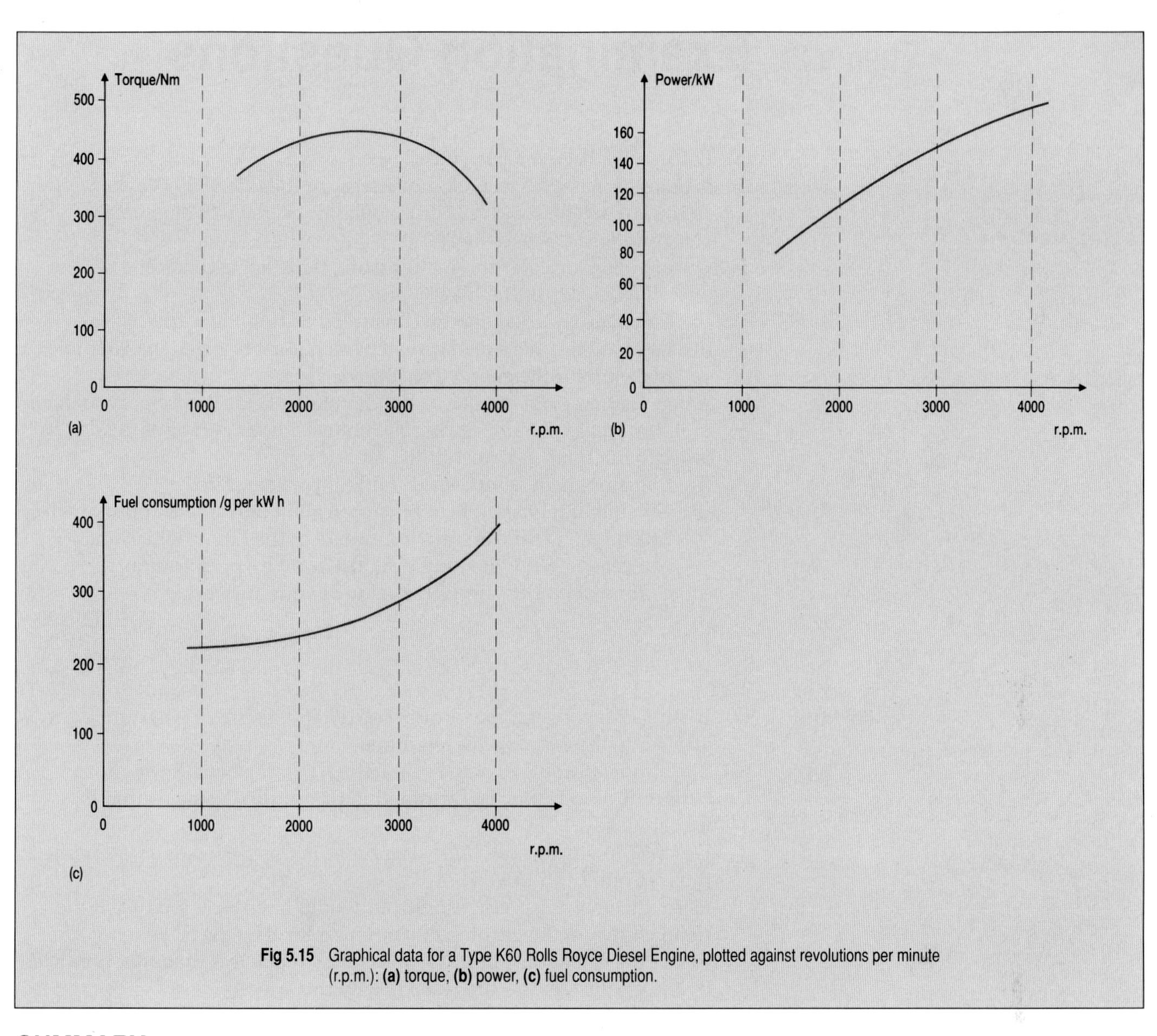

Fig 5.15 Graphical data for a Type K60 Rolls Royce Diesel Engine, plotted against revolutions per minute (r.p.m.): **(a)** torque, **(b)** power, **(c)** fuel consumption.

SUMMARY

Heat engines use a cycle of changes in order to convert heat into work. These changes are conveniently shown on indicator diagrams which, for a gas, are drawn to show how the pressure changes with the volume.

The efficiency of a heat engine is given by

$$\text{efficiency} = \frac{\text{net work output}}{\text{total heat input}}$$

$$= \frac{\text{heat input} - \text{heat output}}{\text{heat input}}$$

There is an upper limit to the efficiency given by

$$\text{maximum efficiency} = 1 - \frac{\text{low temperature}}{\text{high temperature}}$$

A heat pump is a heat engine working in reverse. Its coefficient of performance is given by

$$\text{coefficient of performance} = \frac{\text{heat supplied to high temperature}}{\text{work done}}$$

Theme 2: **Examination Questions**

T2.1

(a) Distinguish between *energy* and *power*.

(b) A hiker of mass 80 kg consumes energy at the rate of 80 W when resting. When walking on a horizontal track, the rate of energy consumption increases by 250 W.
During one particular three-hour walk, the hiker ascends 500 m. The overall efficiency of the hiker's muscles is 12%.
(i) Calculate the total energy consumed during the walk.
(ii) Estimate the mass of chocolate which must be eaten in order to make up for the energy consumed.

(c) When visiting a hot country, a holidaymaker finds it necessary to drink an extra 1.6 kg of water during the hottest 5 hour period of the day in order to maintain a correct fluid balance.
(i) Explain why this increased fluid intake is required.
(ii) Given that the latent heat of vaporisation of water is approximately 2400 kJ kg^{-1}, make an upper estimate of the rate at which energy could have been removed from the body by perspiration.
(iii) Comment on the numerical value of your answer.

(UCLES 1990)

T2.2

(a) Distinguish between the terms *resources* and *reserves* as they are applied to the existence of fossil fuels.

(b) The annual output, C, of the coal industry in the United Kingdom between 1867 an 1907 is illustrated in the graph. C is measured in megatonnes.
Describe in words how coal output rose in this period and determine the coal output in 1870.
Estimate what the predicted output would have been for 1970 if the trend shown in the graph had continued for the next 60 years.
In fact the actual output in 1970 was very much less than that predicted by the graph. Discuss the reasons why this was so.

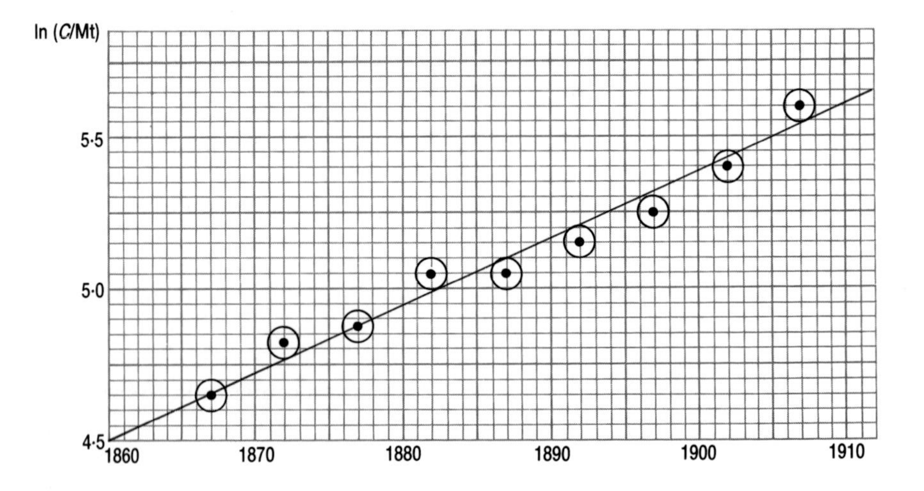

(c) In the nineteenth century, coal was the principal fuel used to provide domestic hot water. Suggest how, in a sunny climate, solar power might be used to provide hot water for a small isolated community of about 20 people. Your answer should include simple diagrams and, wherever possible, calculations for which you may have to make estimates as well as make use of the following data:

solar flux at Earth's surface at midday	≈ 1 kW m^{-2}
specific heat capacity of water	≈ 4 kJ kg^{-1} K^{-1}
average volume of hot water needed per day per person	≈ 10 litres

State any assumptions made in arriving at your estimates.

(ULSEB 1990)

T2.3

This question is about energy dissipation by the human body.
The table gives the rates of thermal energy production in the human body when resting and when working hard. The third column shows the corresponding rates of rise of body temperature assuming that no energy is lost from the body. The values of the required rates of evaporation in the final column are calculated assuming that the body temperature is kept constant, with all the thermal energy being removed by the evaporation of perspiration.

Activity	Rate of production of thermal energy /W	Rate of rise of body temperature /K h^{-1}	Rate of evaporation of water required to maintain constant temperature /kg h^{-1}
Resting	75	1.0	0.11
Hard work	1200	16.0	1.80

Other data and information

Mean mass of the human body: 65 kg
Marathon race distance: 42 km
Mean body temperature before starting the Marathon race: 37 °C (310 K)
Maximum possible rate of energy dissipation by evaporation from the surface of the body: 650 W
Human body ceases to function normally at 41 °C.
Severe dehydration occurs if more than 5% of body mass is lost by evaporation.
Specific heat capacity of water: 4200 J kg^{-1} K^{-1}
Mean specific heat capacity of the human body is approximately equal to that of water.
Mean specific latent heat of vaporisation of perspiration (water) between 35 °C (308 K) and 45 °C (318 K): 2.4×10^6 J kg^{-1}

(a) (i) Deduce the rate of rise of temperature per kilowatt of thermal power produced assuming that no energy is lost from the body.
(ii) Calculate the corresponding rate of evaporation of perspiration per kilowatt of thermal power produced assuming that the body remains at a constant temperature.

(b) A runner in a marathon race produces thermal energy at a rate of 900 W and completes the race in 2.4 h.
(i) Assuming that the runner loses energy only by evaporation, calculate the minimum rate of rise of temperature of the body.
(ii) Use your answer to (i) to explain why there must be some other mechanism for removing energy from the body.
(iii) The only other important mechanism for dissipating thermal energy from the body occurs in respiration when water is vaporised in the lungs and exhaled. Calculate the minimum mass of water vapour which the runner would need to exhale during the race if the body temperature is not to rise above 41 °C.

(iv) Calculate the total water loss from the runner's body during the race. Comment on the possible danger of dehydration for the runner.

(UCLES specimen 1990)

T2.4

(a) The total reserve of geothermal energy is estimated to be 4.1×10^{25} J. Of this resource, about 2% is believed to be hot enough for use in electricity generation of which only about one fifth is thought to be recoverable by existing technology. The conversion efficiency would be about 2.5%.

Current world primary energy consumption is at a rate of about 2×10^{20} J per year of which about one quarter is used for electricity generation.

(i) Explain the meaning of the terms *primary energy* and *geothermal energy*. Suggest two physical phenomena which act as sources of geothermal energy.

(ii) Demonstrate whether the figures suggest that geothermal energy could meet a significant proportion of the world's demands for electrical energy.

(b) (i) Draw a labelled block diagram of a conventional power station using a fossil fuel. Discuss the need for the steam to be heated to a high temperature. (You should make reference to the second law of thermodynamics.)

(ii) Water from geothermal sources generally has a temperature between 75 °C and 200 °C. It is often rich in minerals, including sulphur. Suggest how this water might be used to reduce the fuel consumption of a conventional power station. Suggest one other commercial use for water from geothermal sources.

Discuss briefly the environmental implications of the use of geothermal energy.

(ULSEB 1989)

Theme **3**

THE PHYSICS OF TRANSPORT

The rapid development of physics during the 18th century was followed by a rapid development of transport during the 19th and 20th centuries. This was no coincidence.

Once a good understanding of physical principles was established, use could be made of those principles, particularly in new modes of transport. This is why a study of energy can be illustrated well by consideration of different modes of transport and of heat engines.

The final theme in this book does just this. Its aim is to enable you to see physics in action in common modes of transport encountered in everyday life.

The development of the car in this century has resulted in some changes in shape in order to improve efficiency.

Chapter 6

FORCE

In this chapter some of the modes of transport with which we are all familiar are analysed using Newton's laws of motion.

> **LEARNING OBJECTIVES**
>
> After studying this chapter you should be able to:
>
> 1. apply Newton's laws to many different forms of transport;
> 2. explain how the motive force is generated in different examples;
> 3. understand the application of the Bernoulli effect;
> 4. use Archimedes' principle.

6.1 NEWTON'S LAWS

An understanding of Newton's laws of motion is essential when studying the physics of transport. All forms of transport involve motion and Newton's laws are the principles on which a study of motion is based.

Formal statements of Newton's laws are:

First Law: Every body continues in its state of rest or state of uniform motion in a straight line unless acted upon by a force.

Second Law: The rate of change of momentum of a body is directly proportional to the resultant force on the body.

Third Law: If body A exerts a force on body B then body B exerts an equal and opposite force of the same type on body A.

For a body of fixed mass the first two laws are summarised by the equation:

$$\underset{\text{in newtons}}{\text{force}} = \underset{\text{kilograms}}{\text{mass in}} \times \underset{\text{in metre second}^{-2}}{\text{acceleration}}$$

The units need to be stated here to ensure a valid equation. If any other units are used then a dimensionless constant may be needed in the equation to give correct numerical values.

The implication of the first two laws is that force causes acceleration. If zero resultant force acts on a body it will have no acceleration; if there is a resultant force on a body then it will be accelerating in the direction of the resultant force. Note that the force which controls the acceleration of the body is the resultant force. This is the sum of all the individual forces acting. A body which has no acceleration has zero resultant force acting on it. This does not mean that there are no forces at all acting on a body travelling with constant velocity, but that when all the forces which are acting are added together then the result is zero. This is an extremely important idea in all forms of transport. An aeroplane travelling with constant velocity at its cruising altitude is not accelerating, so the lift on it must be equal and opposite to its weight and the drag on it must be equal and opposite to the thrust.

Newton's Third Law can be used to prove the Law of Conservation of Momentum. The law has no known exceptions. It applies at all times with bodies which are travelling at constant velocity or accelerating: it applies on a cosmic scale and with sub-atomic particles. Einstein's theory of relativity incorporates the law within it.

At first sight the Third Law appears to state that forces always balance out to make acceleration impossible, but this is not so because the forces being discussed are acting on different objects. If you throw a ball vertically upwards then you exert an upwards force P on the ball which is greater than the weight of the ball. The ball accelerates while you exert force P. During this time the ball exerts a force P of equal magnitude but opposite direction on you. This is one of several forces acting on your hand while you are throwing the ball. If all the forces acting on your hand are added together then they will be found to give a resultant which causes the upward acceleration of your hand. The value of the forces may change from moment to moment and consequently the acceleration will not be constant, but at all times the force which the ball exerts on the hand is equal and opposite to the force which the hand exerts on the ball. Now consider a practical problem with some typical numerical values.

A lorry of mass 8000 kg is towing a trailer of mass 20 000 kg. The horizontal drag on the lorry is 6000 N and on the trailer is 8000 N. For an acceleration of 1.3 m s^{-2} find (a) the horizontal force P the lorry exerts on the trailer, (b) the horizontal force Q the trailer exerts on the lorry and (c) the driving force F which the road exerts on the lorry.

Do not be tempted to draw only one diagram for this type of problem. Fig 6.1(a) is a diagram showing the horizontal forces acting on the lorry and Fig 6.1(b) is the corresponding diagram for the horizontal forces acting on the trailer.

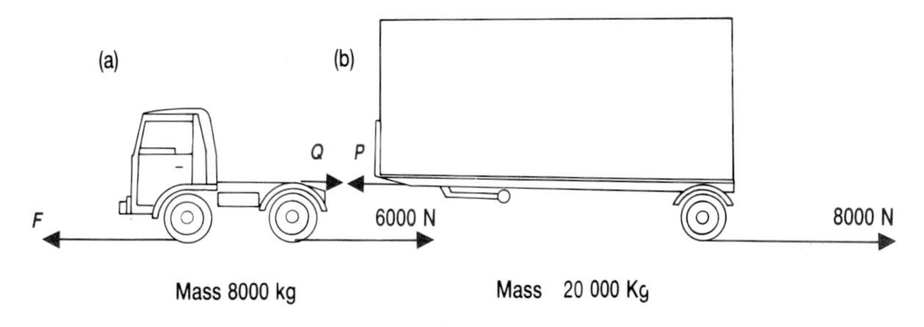

Fig 6.1 Free body force diagrams showing the horizontal forces acting on **(a)** a lorry and **(b)** its trailer.

(a) Consider first the trailer. Since P acts forwards and 8000 N drag acts backwards the resultant force acting on the trailer is $(P - 8000)$ N forwards.

force	= mass	× acceleration
in newtons	in kilograms	in metre second^{-2}

$$(P - 8000) = 20\,000 \times 1.3$$
$$P = 26\,000 + 8000$$
$$= 34\,000 \text{ N forwards}$$

(b) By Newton's Third Law the force the trailer exerts on the lorry is equal and opposite to the force the lorry exerts on the trailer, so

$$Q = 34\,000 \text{ N backwards}$$

(c) Resultant force acting on lorry = (F – 34 000 – 6000) N forwards

force	= mass	× acceleration
in newtons	in kilograms	in metre second^{-2}

$$(F - 34\,000 - 6000) = 8000 \times 1.3$$

$$F = 10\,400 + 40\,000$$
$$= 50\,400 \text{ N forwards}$$
$$= 50\,000 \text{ N to 2 sig figs.}$$

Force

Central to Newton's laws is the idea of **force**. Force is a term with which everyone is familiar and which seems to pose no difficulty, but all is not as simple as it seems at first sight. Frequently when the word *force* is required the word *pressure* is used, incorrectly. 'A lorry which causes a pressure of 42 tonnes would do a great deal of damage to our roads and buildings' is an incorrect use of the word pressure. Pressure implies force per unit area: here no mention is made of area and to be scientifically correct the word *force* should be used instead of pressure.

Even when the term is used correctly there are still problems to be resolved. For single forces these frequently centre on the direction in which a force acts and on its point of application. For more complex systems, involving many forces, problems lie in the complexity of the situation and the consequent need to be certain that all the forces acting are taken into account and that no duplication occurs. It helps to ensure accuracy if, whenever a problem is to be analysed, the object under investigation is clearly stated and a force diagram of that object, and that object only, is drawn. There will be a **contact force** exerted on an object wherever it touches anything else and there will also be the weight of the object to show on the force diagram. It makes a good check on a force diagram to count round the diagram to see that you have a force shown at every point of contact plus one force for the weight. In the example of the lorry towing a trailer two force diagrams were needed, one of the trailer and one of the lorry. By separating the two parts it is more likely that the direction of the forces will be correct. It also helps to label forces more clearly than is frequently done. A label 'P' says very little about a force. A label 'force caravan exerts on car' gives a good deal more information. It must be on a diagram of the car and it implies that somewhere on another diagram there must be an equal and opposite force which the car exerts on the caravan.

Friction

When one object touches another they exert forces on each other. The directions in which the forces act will always be opposite to one another. Sometimes the directions are at right angles to the tangent at the surfaces where contact is made but this is not necessarily so. Fig 6.2(a) shows the heel of a person's shoe exerting a force vertically downwards on a path. (A gap is left between the heel and the path so that you can see where the forces act.) The force can be at an angle to the vertical, however, as shown in Fig 6.2(b). Here the person is pushing downwards and forwards on the path and so, using Newton's Third Law, we see that the path is exerting an upwards and backwards force on the heel. This force will have the effect of decelerating the foot. If the heel had been hitting a slippery surface then the backward component of the force would have been less and so there would have been a smaller force decelerating the foot's forward movement and slipping would have been more likely.

The reverse process is occurring in Fig 6.2(c). Here the person is pushing the sole so that it exerts a downward and backward force on the path. The upward and forward force the path exerts on the person is, in part, the force which gives the person an acceleration. The horizontal component of the force acting on the shoe is called the **force of friction**. This is shown as F on Fig 6.2(c). The vertical component of the force is called the **normal contact force** C. Note that none of these forces is the weight of the person. When walking, the **force of friction** is sometimes forward, sometimes backward and sometimes zero. The size of the force can be anything from zero to a value several times the weight of the person. A baby learning to walk is actually learning to control these forces and to make them suitable to take her where she wants to go. Once walking is accomplished this very complex task is performed by most people without thinking about it, but anyone who has damaged legs as a result of accident or disease finds control of the forces in walking very difficult.

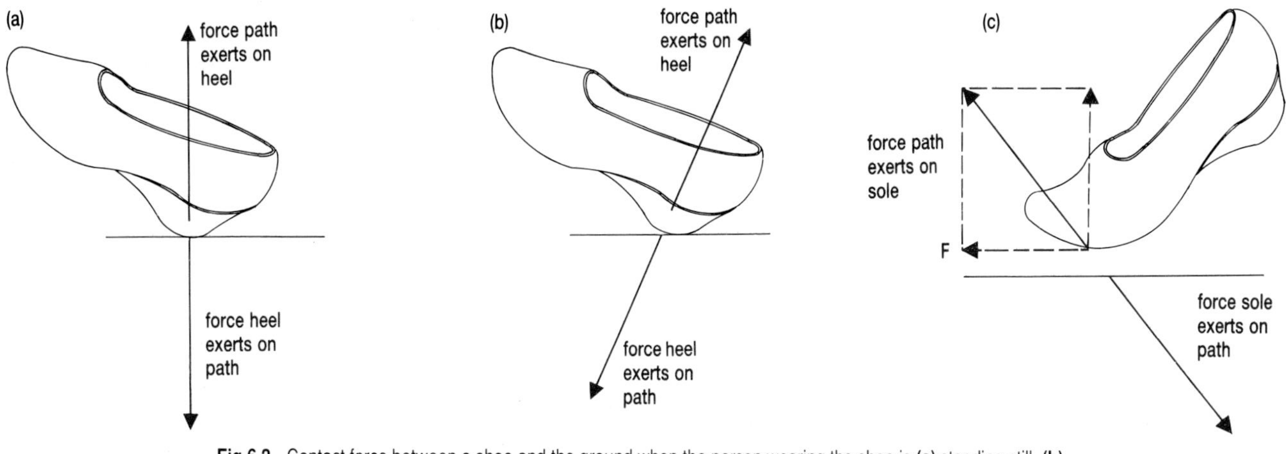

Fig 6.2 Contact force between a shoe and the ground when the person wearing the shoe is **(a)** standing still, **(b)** stopping, **(c)** accelerating.

Friction therefore is a vitally important force. It is friction which allows you to walk and which enables all vehicles to be driven, through their wheels.

Mass

The mass of an object is the property it has which determines how difficult it is to accelerate. The SI unit of mass is the kilogram. Another word for mass is **inertia**. One object which has ten times the mass of another is ten times more difficult to accelerate. This implies that if the same force is applied to both the objects then the one with ten times more mass will only have one-tenth of the acceleration. This has important implications in many aspects of transport as the following calculations show.

An unloaded minibus and its driver have a total mass of 1800 kg and a braking force of 5000 N can be applied to it. What is the minimum distance within which it can stop from a speed of 30 m s^{-1}?

$$\text{Acceleration} = \text{force/mass} = -5000\ \text{N}/1800\ \text{kg} = -2.78\ \text{m s}^{-2}$$

The minus sign indicates that the force is acting in a direction opposite to the direction of the velocity.

$$\text{Since } v^2 = u^2 + 2as$$
$$0 = 30^2 - 2 \times 2.78 \times s$$
$$s = 30^2/(2 \times 2.78) = 162\ \text{m}$$

If the same minibus then carries 15 people and their luggage so that the total mass rises to 3000 kg, find the new stopping distance from a speed of 30 m s^{-1}.

Repeating this question with a mass of 3000 kg in place of 1800 kg gives acceleration = force/mass = –5000N/3000 kg = – 1.67 m s^{-2}

$$\text{Since } v^2 = u^2 + 2as$$
$$0 = 30^2 - 2 \times 1.67 \times s$$
$$s = 30^2/(2 \times 1.67) = 270\ \text{m}$$

In practice the extra load does enable a greater braking force to be applied so that the stopping distance is not increased as much as the example shows, but anyone who regularly drives heavily laden vehicles is well aware of the difficulty of stopping them. The reluctance of an object to stop once it is travelling is its mass. Note that it is *not* a force which is trying to keep the object going.

Weight

The weight of an object is the gravitational force pulling the object towards the centre of the Earth. It is a force acting on the object itself so it is incorrect to talk of the weight of a person acting on the chair he is sitting on. The SI unit of weight must be the newton since weight is a force. Near the surface of the Earth a mass of one kilogram has a weight of approximately 9.8 newtons and this force does not change as a result of any acceleration occurring. If you jump from a wall and land heavily your weight does not increase as you land. Your weight remains constant the whole time; the force of gravity on you does not change just because you make contact with the ground. What does change is the force you exert on the ground and therefore the force the ground exerts on you. It is a surprising fact that you can see the effect of the force of gravity on an object but you cannot feel the force of gravity on yourself. You feel the force with which a chair supports you when you sit in it. This force is not a gravitational force but is a contact force. All contact forces are electrical because all matter is made of charged particles. A contact force is noticeable because it alters your shape. If you jump from a diving board you accelerate towards the water because of the force of gravity but you cannot feel the force acting. Indeed this situation is sometimes, incorrectly, called weightlessness because you cannot feel any weight, and is the same free fall condition that an astronaut is in when orbiting the Earth in a satellite.

Drag

As stated earlier in this section a force of friction between the ground and a moving person can be forwards or zero or backwards. For a forwards force, however, the person must be doing some work and hence must be using some chemical energy. This can be extended to a vehicle. If the frictional force on a car is in the forward direction then this can only come about as a result of the engine working on the driving wheels. For wheels which are not driven, the rolling wheels, the frictional force can only be in a backwards direction and this force is nearly independent of the speed of the car. The force does depend on the type of surface over which the car is travelling and on the total weight of the car. For a car of mass 1000 kg travelling on a typical tarmac surface the resisting force of rolling resistance is likely to be about 150 N and this force constitutes part of the total drag on the car.

Fig 6.3 A four wheel drive vehicle such as this landrover is designed for travelling relatively slowly over rough ground, streamlining it is not important.

Apart from the frictional force at the rolling wheels the other constituent of drag is the force exerted by the air through which the car is travelling. This force is approximately proportional to the square of the speed of the car. The force F_{air} is given by the equation

$$F_{air} = \tfrac{1}{2} C A \rho v^2$$

where A is the frontal area of the car, ρ is the density of air (about 1.2 kg m^{-3} at normal temperatures), v is the speed of the car and C is a dimensionless constant called the **drag coefficient**.

ASSIGNMENT

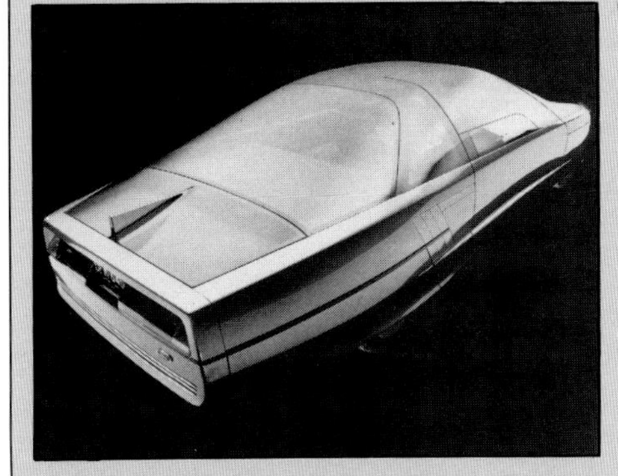

Fig 6.4 A modern car is designed to have as low a drag coefficient as possible. This is what cars might look like in the future.

Cars and air resistance

Car manufacturers take great care to design the shape of a car so that air resistance is kept to a minimum. Cars in 1990 typically have a drag coefficient of 0.30. The Ford Motor Company are aiming to reduce this to 0.25 within five years but state 'this is very difficult to achieve in a mass produced car. It means spending a lot of money on special airflow devices ... and perhaps running on special narrow tyres'. Air resistance by the tyres themselves is a significant part of the total air resistance. Fig 6.4.

A car has a frontal area of 1.9 m^2, has a maximum speed of 40 m s^{-1}, a drag coefficient of 0.30 and a rolling resistance of 150 N.

1. Use the equation for air resistance, $F_{air} = \frac{1}{2}CA\rho v^2$ to find F_{air} at its maximum speed (the density of air is 1.2 kg m^{-3}).
2. Find the total resistive force on the car at its maximum speed.
3. What is the driving force when the car is at its maximum speed?
4. Show that the power the engine is supplying when the car is at its maximum speed is 28 000 W to 2 significant figures.

The output power of the engine must be appreciably greater than this, since power will be lost in the gear box, the transmission system and the driving wheels as well as in auxiliary systems such as the electrical generator. Since the power output from the engine is limited by the laws of thermodynamics to a relatively small fraction of the energy supplied by burning the petrol, the percentage of the input of energy from the fuel which finally drives the car is often only around 20%. A car with the output power required in this example could well have the following power characteristics:

Power supplied by the burning petrol	140 kW
Theoretical maximum mechanical power output	85 kW
Practical output mechanical power	40 kW
Mechanical power lost in driving accessories	3 kW
Mechanical power wasted in transmission system	9 kW
Mechanical power used for propulsion	28 kW

These figures are illustrated in the Sankey diagram, Fig 6.5.

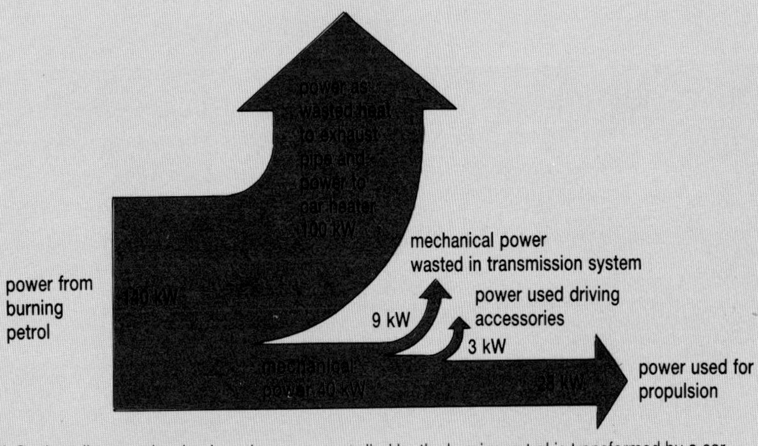

Fig 6.5 A Sankey diagram showing how the energy supplied by the burning petrol is transformed by a car.

Repeat this assignment for a car such as the one which the Ford Motor Company are hoping to produce with a drag coefficient of 0.25. If the power output of the engine remains unchanged, show that the maximum speed of the new car will be 42.4 m s^{-1}.

What other factors, apart from the drag coefficient, could reduce the air resistance on the car? What would be the problem with each of your suggested methods of drag reduction from the point of view of the car designer?

Streamlining

When a solid object travels through a viscous fluid there is a resistive force exerted by the fluid on the object. The size of this force depends upon the speed of movement as shown above, on the viscosity and density of the fluid and upon the frontal area and shape of the object. In the case of a person moving through air at walking speed the force is barely noticeable. The speed is low and the viscosity and density of the air are small. Only if the air itself is moving at higher speed, that is if there is a wind blowing, does the force affect the way the person walks. A good athlete, however, is concerned to reduce viscous drag. A cyclist will wear smooth clothing which fits tightly so that the air flow past the cyclist has minimum resistance. It might be worthwhile designing a bicycle so that the cyclist travels through the air in a horizontal position rather than an upright one. Certainly in vehicles designed for competitions in which the winner is the one that travels furthest on a cupful of petrol the driver is arranged to be in a prone position, Fig 6.6.

Fig 6.6 At the Shell mileage marathon at Silverstone this car is capable of travelling over 300 km on a litre of petrol.

A free-fall parachutist, before opening the parachute, can reach a terminal velocity of around 60 m s^{-1}. At this rate of fall the viscous force exerted by the air is equal and opposite to the parachutist's weight. The figure of 60 m s^{-1} varies considerably according to several factors. The parachutist may be falling in a flat position through the air or in a diving position or curled up into a ball; the density of the air may be higher or lower according to the temperature and the altitude of the parachutist; a heavy parachutist reaches a higher terminal velocity than a light one because the area of a person is not proportional to the person's weight. This last point is important in many scaling problems. Note that, assuming the same shape, a child of height 1 m compared with an adult of height 2 m has one-eighth of the weight and one-quarter of the frontal area. This makes the air resistance on the child a more significant force than it is on the adult. The child would therefore have a smaller terminal velocity when falling than the adult.

INVESTIGATION

The effect of speed on air resistance

Experiments of this kind are not usually done by using an object moving through the air but rather by use of a wind tunnel in which the air is moved past a stationary object. In the absence, in a school laboratory, of elaborate wind tunnel test facilities an alternative method has to be devised. The method suggested is simple and the principle is that a stream of air is blown vertically downwards on to the pan of a sensitive top pan balance. The reading on the balance can easily be converted into a reading of the force exerted in newtons. The problem of establishing the average speed at which air in the stream is moving is overcome by using a known volume of water to push the air through the jet being used. Some trial runs may be necessary to establish a suitable size of jet. The suggested apparatus is shown in Fig 6.7 using a jet size of area of cross section about 2 mm^2.

Vary the rate at which air passes through the jet and measure for each different flow rate the force exerted on the balance. The readings taken should include:

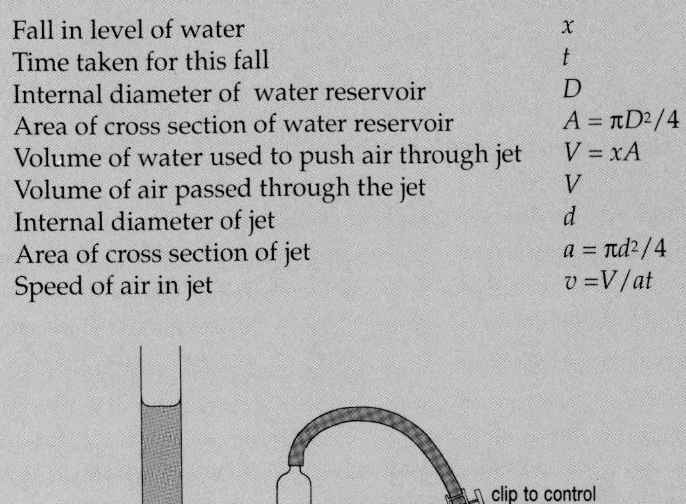

Fall in level of water	x
Time taken for this fall	t
Internal diameter of water reservoir	D
Area of cross section of water reservoir	$A = \pi D^2/4$
Volume of water used to push air through jet	$V = xA$
Volume of air passed through the jet	V
Internal diameter of jet	d
Area of cross section of jet	$a = \pi d^2/4$
Speed of air in jet	$v = V/at$

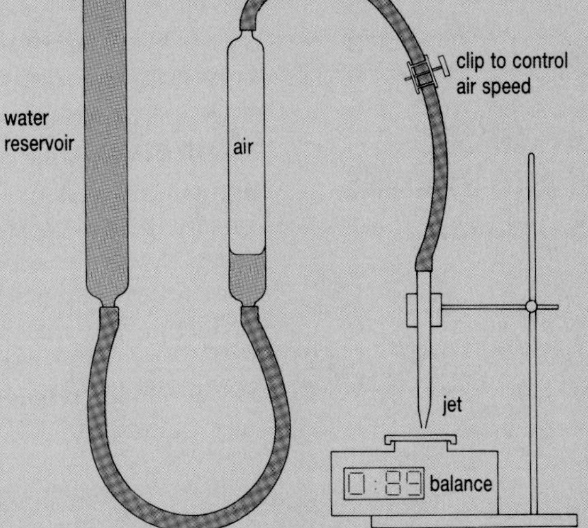

Fig 6.7 The arrangement suggested for measuring how the force exerted by a moving air stream varies with the speed of the air stream.

Now plot a graph of force exerted against the square of the speed of air through the jet.

QUESTIONS

6.1 A car of mass 900 kg is towing a caravan of mass 400 kg. The force which the car exerts on the caravan through the tow bar is a horizontal forward force of 500 N and the caravan has a forward acceleration of 0.80 m s^{-2}. Find

(a) the resultant force on the caravan;

(b) the horizontal drag on the caravan;

(c) the acceleration of the car;

(d) the resultant force on the car;

(e) the driving force on the car if the horizontal drag on the car due to friction is 200 N;

(f) the resultant force on the car and caravan, regarded as a single object.

6.2 Draw force diagrams showing and labelling all the forces caused by gravity and by contact on the following pairs of objects.

(a) a book at rest on a table and the table,

(b) a person in a lift and the lift when the lift is

(i) stationary

(ii) rising at constant velocity
(iii) falling at constant velocity
(iv) rising and accelerating
(v) rising and slowing down
(vi) falling and slowing down.

6.3 A train driver in a train of mass 4.0×10^5 kg and travelling at 60 m s^{-1} needs to apply his brakes, with an assumed constant force, 120 s before he stops. What is the deceleration of the train? What force is applied and how far does the train travel while braking? Use your answer to suggest where to place the automatic control switch for level crossing barriers. What problem(s) may arise with automatic barrier control?

6.4 A car, of mass 800 kg, has an output power of 60 kW and a top speed of 40 m s^{-1}. What drive force does it supply when travelling at its top speed? If it uses this drive force when the resultant drag on it is only 1100 N what will then be its acceleration?

6.2 GENERATION OF DRIVING AND BRAKING FORCES

Momentum

Momentum is defined as the product of the mass of an object and its velocity. It is a vector whose direction is the same as the velocity of the object. Newton's Second Law of Motion states that the rate of change of the momentum of an object is proportional to the resultant force on the object.

Algebraically this gives:

$$\frac{dp}{dt} \propto F \text{ where } p \text{ is the momentum.}$$

If the force is constant this can be written:

$$\frac{\text{change in momentum}}{\text{time interval}} \propto \text{force}$$

or

$$\frac{\text{change in mass} \times \text{velocity}}{\text{time interval}} \propto \text{force}$$

Since the mass of an object is often constant this gives:

$$\text{mass} \times \frac{\text{change in velocity}}{\text{time interval}} \propto \text{force}$$

Since the change in velocity per unit time is the acceleration we get:

$$\text{mass} \times \text{acceleration} \propto \text{force}$$

If the mass is in kilograms, the acceleration in metre second^{-2} and the force in newtons this gives the familiar equation

$$\text{force} = \text{mass} \times \text{acceleration}$$

The constant of proportionality is made to be equal to 1 by defining the newton as that force which gives a mass of 1 kilogram an acceleration of 1 m s^{-2}.

Apart from the need to be thoroughly familiar with this equation the other reason for repeating it here is to emphasise that a resultant force on an object causes its momentum to change. When any mode of transport starts, therefore, its motive power unit must be arranged in such a way that an *external* force is applied to the vehicle. Internal forces are useless because they do not produce a resultant force acting on the vehicle. You cannot get a reluctant car to start by pushing it from the inside; you have to get out and push it from outside.

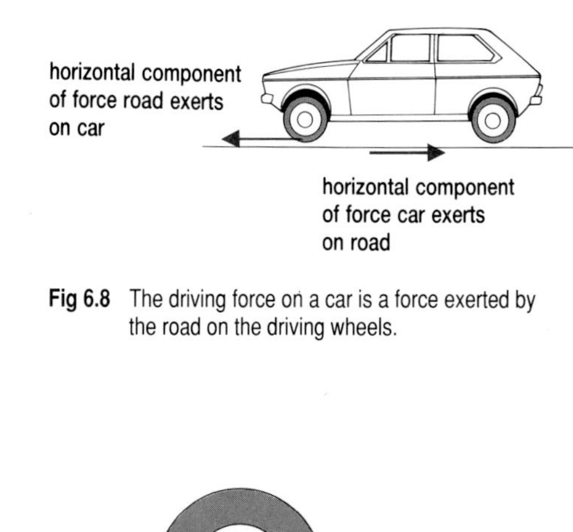

Fig 6.8 The driving force on a car is a force exerted by the road on the driving wheels.

Action of wheels

How then does an internal engine manage to provide an external force on a vehicle? The answer of course is because it drives the wheels and the wheels exert a force backwards on the road. The road consequently exerts a forward force on the vehicle (Newton's Third Law). As an example, consider a front wheel drive car. It is not sensible to show drive forces on the car coming from the engine itself. Instead show the force acting forward on the front wheels, Fig 6.8.

One of the front wheels of the car is shown in Fig. 6.9. The drive shaft from the engine exerts a torque T on the front wheel. If the car is not accelerating then this torque must be balanced by the moment of the force which the road exerts on the tyre. If the torque exerted by the drive shaft is 200 N m and the car is not accelerating, the driving force exerted by a wheel of radius 25 cm can be found.

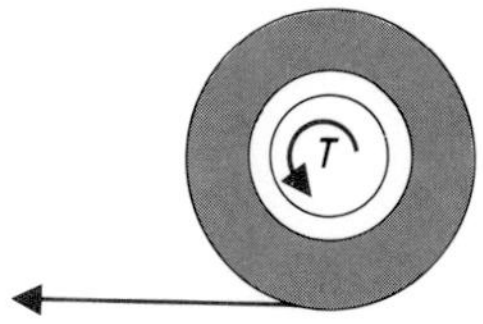

T

torque *T* from drive shaft

horizontal component of force road exerts on car (F)

Fig 6.9 The drive force is generated by a torque which the axle exerts on the wheel.

$$\text{torque supplied} = \text{moment of } F$$

Provided these two quantities are the same, the resultant torque on the wheel is zero and the wheel will not undergo rotational acceleration. It will therefore go round at a steady speed.

So,

$$200 \text{ N m} = F \times 0.25 \text{ m}$$

$$F = \frac{200 \text{ N m}}{0.25 \text{ m}} = 800 \text{ N}$$

If now the driver allows more petrol/air mixture to reach the engine by pressing on the accelerator pedal, the power output from the engine increases with a consequent increase in the output torque. The torque then becomes larger than the moment of the force exerted by the road and this results in an acceleration of the rotation of the wheel and hence an acceleration of the car itself. This acceleration will continue until less petrol/air mixture is supplied or until the extra speed of the car causes increased air resistance.

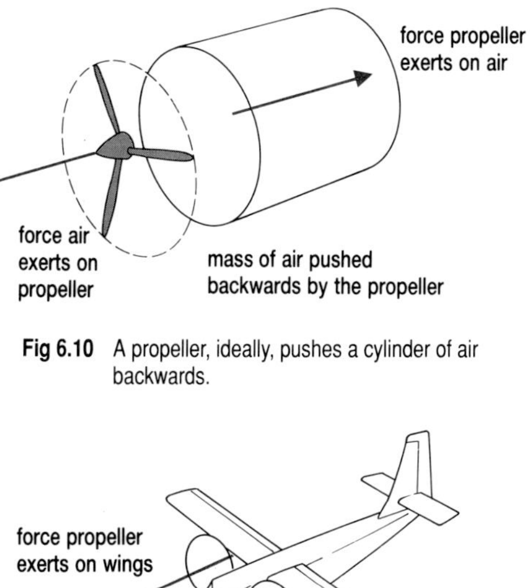

Fig 6.10 A propeller, ideally, pushes a cylinder of air backwards.

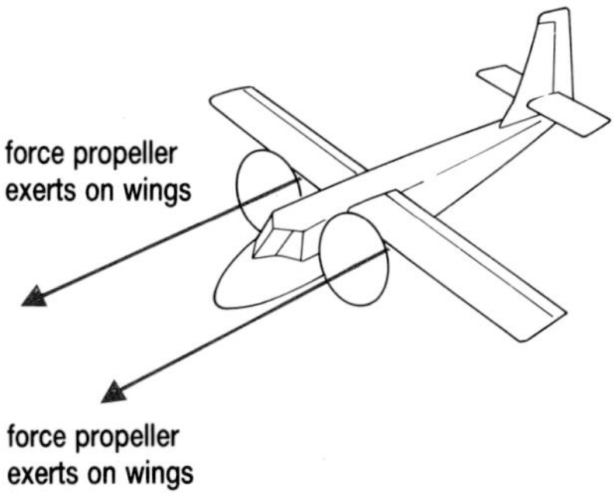

Fig 6.11 A propeller-driven aeroplane is pulled forward by the force which the propellers exert on the aeroplane.

Jet engines and propellers

The driving force in a car originates because the wheel exerts a backward force on a solid, the road. By contrast, both jet engines and propellers exert a backward force on a fluid. The fluid in the case of jet engines is a gas. The jet engine draws in air at the front and the exhaust gas is ejected at high speed from the back. For propeller driven vehicles the fluid is air whereas for propeller driven boats, it is water. The principle for all these systems is the same. Fluid at low velocity in front of the propeller has its velocity increased by the propeller to provide a backward moving mass of fluid. The rate of change of momentum of the fluid is equal to the backward force which the propeller exerts on the fluid. If there is a backward force exerted by the propeller on the fluid then, using Newton's Third Law, there will be an equal and opposite force exerted by the fluid on the propeller. This will be a forward force as shown in Fig 6.10. A chain of forces transfers this force to all the parts of the aeroplane. Since the propeller is fixed to the aeroplane, it cannot move forward without the aeroplane. The plane exerts a backward force on the propeller and the propeller exerts a forward force on the aeroplane's wings as shown in Fig 6.11.

It is worthwhile to consider a particular propeller at this stage. A microlight aeroplane is virtually a hang glider with an engine attached. If the propeller blade is 0.80 m long and the speed of the air is increased by the propeller by 20 m s^{-1} then the driving force can be determined.

Assume that all the air passing through the propeller has a density of 1.25 kg m^{-3} and has its speed increased by 20 m s^{-1}. The propeller causes a cylinder of air to travel at a speed of 20 m s^{-1} (Fig 6.12).

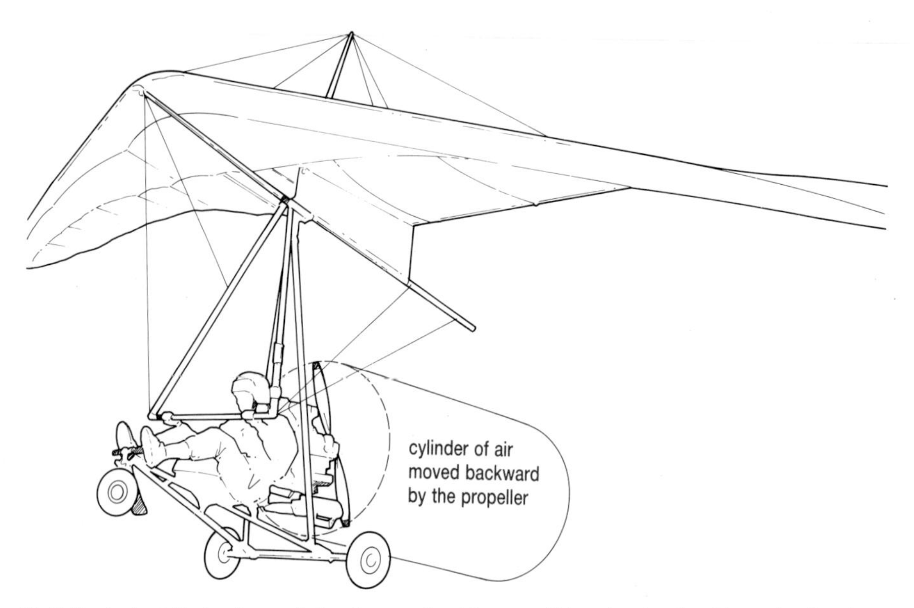

Fig 6.12 A microlight aircraft normally has the propeller at the rear which pushes the aircraft forward as it pushes a cylinder of air backwards.

Volume of air moved per unit time $= 20 \times \pi r^2$

$= 20 \times \pi \times 0.80^2 = 40 \text{ m}^3$

Mass of air moved per unit time $= 40 \times 1.25 = 50$ kg

Momentum transferred to air per unit time $= 50 \text{ kg} \times 20 \text{ m s}^{-1} = 1000 \text{ kg m s}^{-1}$

However, this is the rate of change of momentum. It is therefore the force acting on the air and will also be the driving force on the microlight, that is:

driving force = 1000 N

Propellers might have been used for the propulsion of cars, but wheels are simpler. However, aeroplanes on take-off do use propellers to provide their driving force. It would add too much to the weight of an aeroplane to have its wheels power driven for take off, so in that case the propellers, or jet engines, are used in the way described both on land and when in the air.

Rockets

A rocket motor uses the same principle as a jet engine, but does not draw in any air in the first place. This is why a rocket motor is able to operate in space. A rocket has to carry not only its fuel but also its oxygen for combustion. Often the oxygen is carried in the liquid phase and it is mixed with the fuel just prior to burning. If a rocket motor supplies a constant force to a rocket the acceleration of the rocket will increase because the total mass of the rocket falls as fuel and oxygen are burnt. This is noticeable with a firework rocket. The characteristic whoosh of a firework is due to the acceleration building up very rapidly as the mass of the fuel in the rocket drops. In a space rocket the rate of burning of fuel with oxygen is very fast so that a rocket normally accelerates under power for a comparatively short time (less then 5 minutes) and then coasts with its engines turned off. When it wants to carry out a manoeuvre it will either use small thruster

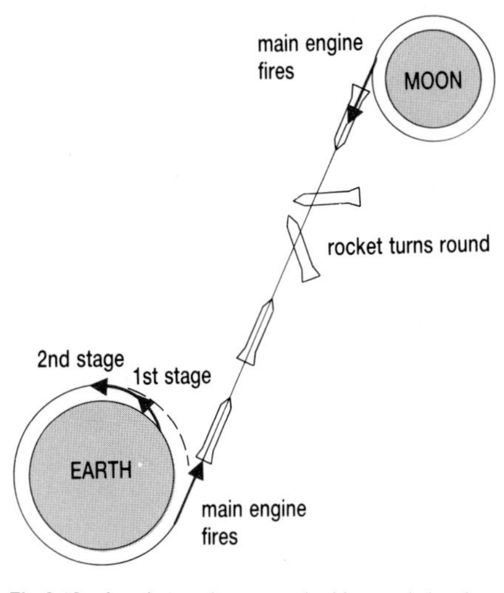

Fig 6.13 A rocket on the way to the Moon only has its engines on for a small fraction of the total time of the journey.

rockets to change course slightly or will switch on its rockets for another burn. Fig 6.13 shows how, on a journey to the Moon, the rocket motors are off for most of the time. After two rocket stages have moved the lunar voyager into orbit it fires its main engine to launch it on its way. As it approaches the Moon it turns round, using thruster motors, and then fires its main engine again so that it slows down and goes into Moon orbit.

Propeller pitch

When a propeller rotates it slices through some air and pushes it backwards. The speed with which the air is pushed depends on several factors: the rate of rotation of the propeller, the speed of the propeller, which increases towards its tip, and the angle at which the propeller is set. A propeller is normally designed with an aerofoil cross section, so that it moves smoothly through the air, Fig 6.14 .

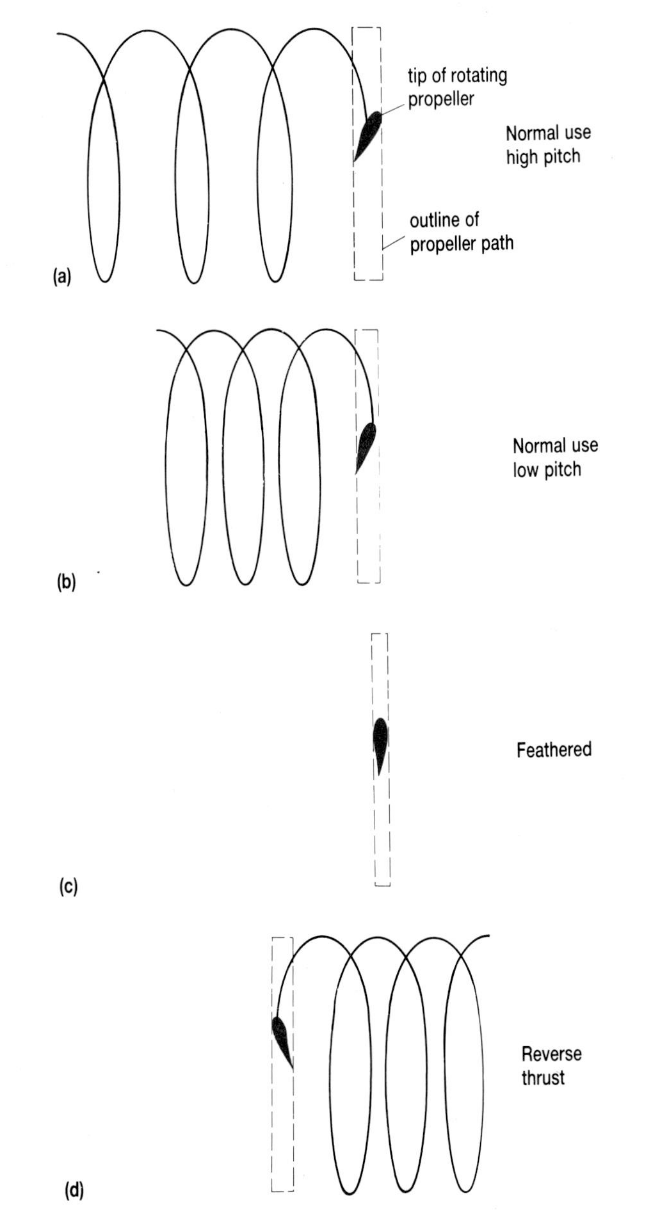

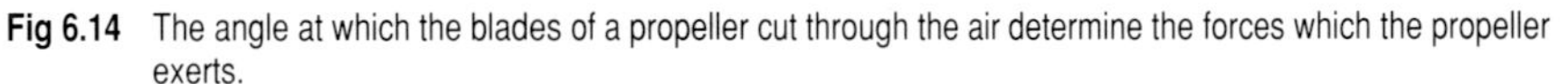
Fig 6.14 The angle at which the blades of a propeller cut through the air determine the forces which the propeller exerts.

The design of a propeller is complex. It has to take into account the different speeds of the blade at different distances from the axis and it has to be shaped to give minimum turbulence. The more turbulence it causes, the greater the drag.

Fig 6.14 also shows how the pitch of a propeller affects the air through which it is travelling. In Fig 6.14(a) the propeller, whose tip only is shown, is slicing large volumes of air at each rotation and pushing them backwards in a helical pattern. This propeller has a large **angle of pitch**. In Fig 6.14(b) the angle of pitch is lower so the air is forced back less quickly. By reducing the angle of pitch to zero, Fig 6.14(c), the propeller ceases to give a forward thrust. This is known as **feathering**. If the angle of pitch is made negative then air is forced in the opposite direction to give reverse thrust.

If this diagram is held sideways then it should be clear in (a) and (b) how the rotor on a helicopter pushes air downwards in a helical pattern. No lift would be supplied by a helicopter with feathered rotors. A helicopter pilot controls the rise and fall of a helicopter by altering the angle of pitch of the rotor blades rather than by altering its rate of rotation. A change in the angle of pitch gives a more immediate response to the pilot.

There is a limit to the angle of pitch. If it is made too large turbulence is set up and a situation similar to that of an aeroplane stalling occurs.

QUESTIONS

6.5 A helicopter has a mass of 10 000 kg and can be considered to direct a cylindrical column of air vertically downwards with a uniform speed of 30 m s^{-1}. What must be the length of each blade in its rotor? The density of air at s.t.p. is 1.3 kg m^{-3}.
Why is a helicopter's load carrying capacity reduced at
(a) high temperatures;
(b) high altitudes?

6.6 A ship's propeller moves water backwards with a velocity of 5 m s^{-1}. What diameter propeller is required if the driving force on the ship is to be 100 000 N? The density of water is 1000 kg m^{-3}.
What essential fact makes the propeller in this question have a smaller diameter than the one in the previous question despite the fact that the force exerted is the same and the water moves much more slowly than the air?

6.7 A jet engine takes in 400 m^3 of air travelling with a velocity (relative to the engine) of 250 m s^{-1}. It ejects the air with a velocity of 900 m s^{-1} backwards. The density of the air at the operating altitude is 0.93 kg m^{-3}. Find the thrust of the engine.

6.3 THE BERNOULLI EFFECT

If a garden hose has a tap on it near the nozzle and it is gripped firmly just before the water comes through the nozzle an interesting effect is noticed. If the water is flowing freely through a fully open tap then the hose seems limp. If the tap is gradually shut the hose swells until with no water travelling through the hose at all the pressure inside reaches its maximum value and the hose is very stiff.

This is an example of the Bernoulli effect in operation. The faster a fluid flows the lower the pressure of the fluid. This seems at first sight to be the opposite of what is expected but it is an effect which can be deduced from the fact that when a fluid speeds up it needs a force to accelerate it. This is provided by the difference in pressure between the high pressure where the flow is slow and the low pressure where the flow is fast. The effect can be noticed when a high-sided lorry overtakes a car towing a caravan. The high speed air in the space between the two vehicles has low pressure. Since the caravan has air at atmospheric pressure on one side of it and low pressure on the side by the lorry there will be a force on the caravan pushing it towards the lorry. Since this is very disconcerting for the drivers of cars towing caravans the towing link between car and caravan often has a stabiliser unit built into it to prevent the caravan from twisting on its towing bracket.

The same effect makes it difficult for naval ships to be provisioned at sea. If a fleet auxilliary steams alongside a ship which it needs to supply, the water between the two ships is moving relative to them, and because of the Bernoulli effect, there is a reduced water pressure between the ships. This results in there being a sideways force on both ships pushing them together. If they stop to reduce this effect then, because they are not under way, they have no rudder control so find it difficult to remain in a fixed position. It is a test of good seamanship to be able to stay a fixed distance away from another ship, perhaps for a couple of hours, while goods are transported across the gap between the ships on a cable.

INVESTIGATION

The Bernoulli effect

Several practical demonstrations of the Bernoulli effect are shown in Fig 6.15. These can be set up quite simply using a stream of air from the outlet of a vacuum cleaner. They are fascinating and it is well worthwhile to set them up. Use the hose from the vacuum cleaner's outlet and fix into it a rubber bung fitted with a piece of glass tubing as shown in Fig 6.15. Some experimenting to find the most suitable diameter glass tubing may be necessary. The air stream can then be directed where it is required. The reduction in pressure which occurs is at the side of the air stream and it is into this region that objects are pulled. In (d) the reduction in pressure can be measured from the difference in water levels inside and outside the tube.

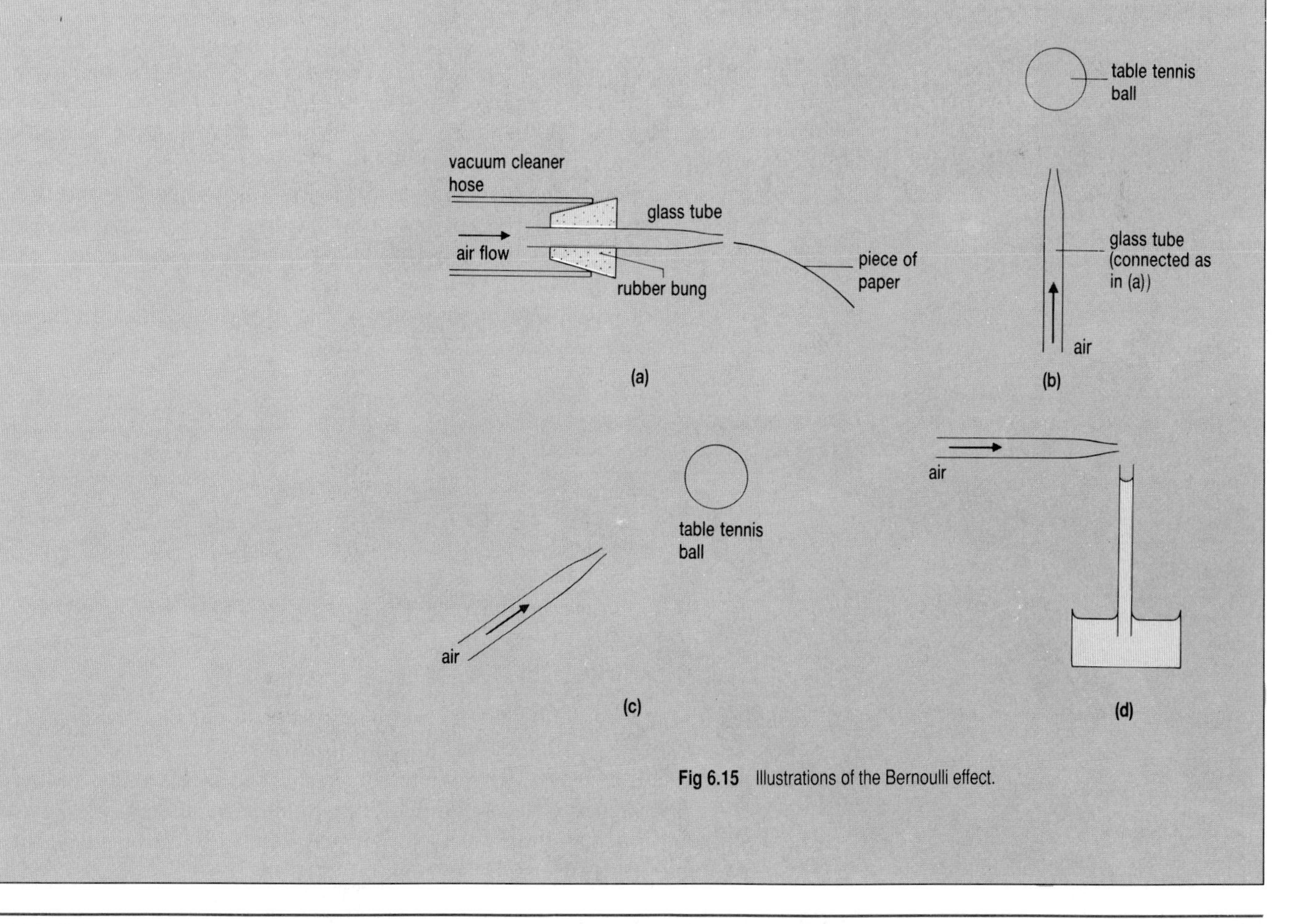

Fig 6.15 Illustrations of the Bernoulli effect.

This principle is used in several practical devices. If the reduction in pressure is sufficient the liquid in the central tube rises into the air stream. In a scent spray the liquid is scent and the air stream then sprays the scent where it is directed. In a carburettor the liquid is petrol and the accelerator pedal in the car controls the rate of air flow past the central tube. Petrol rises up the central tube and evaporates in the air stream to give the required petrol/air mixture which is fed directly into the engine.

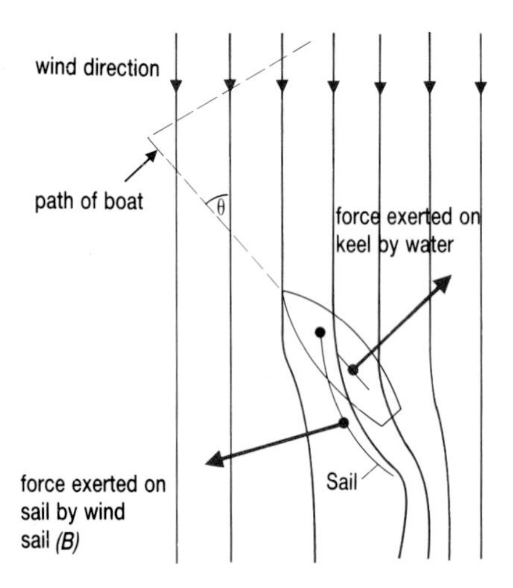

Fig 6.16 When a sailing boat is tacking it is able to use the force which the wind and water exert on it to make headway against the direction of the wind.

The importance of the Bernoulli effect in sailing

A sailing boat is able to make headway when sailing against the wind as a result of using the Bernoulli effect. Fig 6.16 shows a sailing boat tacking, that is, moving towards the direction the wind is coming from by making a zig-zag path through the water. The movement of the wind past the sail results in air slowing down on one side of the sail and speeding up on the other side. This results in a Bernoulli force B on the sail acting from the high pressure, low speed side of the sail to the low pressure, high speed side. If, in addition to this force, the keel of the boat has a force exerted on it by the water, as shown, then the resultant of these two forces can be in the required direction. One of the skills of sailing is to be able to keep the boat as close to the wind as possible without losing the driving force which the Bernoulli effect supplies. On the diagram this means keeping the angle θ as small as possible so that the path of the boat as it tacks uses as few zig-zags as possible.

Lift on an aircraft

The Bernoulli effect also plays an important part in providing **lift** for an aeroplane. Fig 6.17 shows the aerofoil section of a wing and the air flowing past it. The air travelling above the wing has to travel further because of the shape of the wing. It therefore speeds up and, according to Bernoulli, there will be a low pressure region above the aeroplane's wing. The air which passes underneath the wing slows down a little because the wing bottom is at a slight angle, the angle of attack, to the approaching air. This increases the pressure beneath the wing. The increased pressure below, and the reduced pressure above the wing give rise to the lift forces shown in Fig 6.17. For a real situation in which the aeroplane is moving through the air the effect is the same as described here.

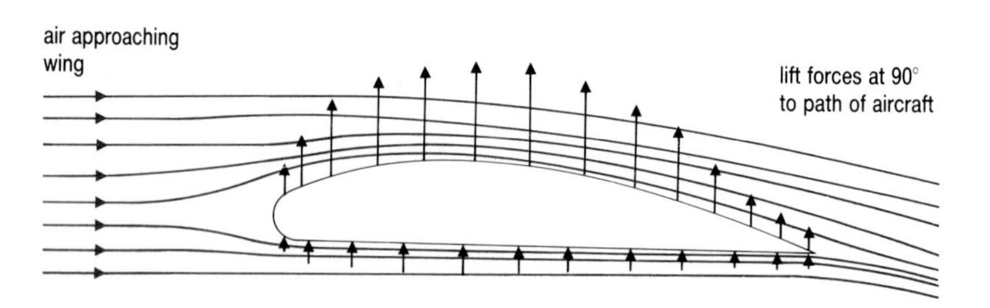

Fig 6.17 Air across an aerofoil section gives rise to lift.

The lift on an aeroplane's wing can be adjusted by varying the **angle of attack**. A low angle of attack gives low lift. If extra lift is required the angle of attack can be increased. However, if the angle of attack is increased to a high value, turbulence is set up and there is a sudden loss of lift. This is called **stalling** and it is most likely to happen when the air speed is low. This is shown in Fig 6.18.

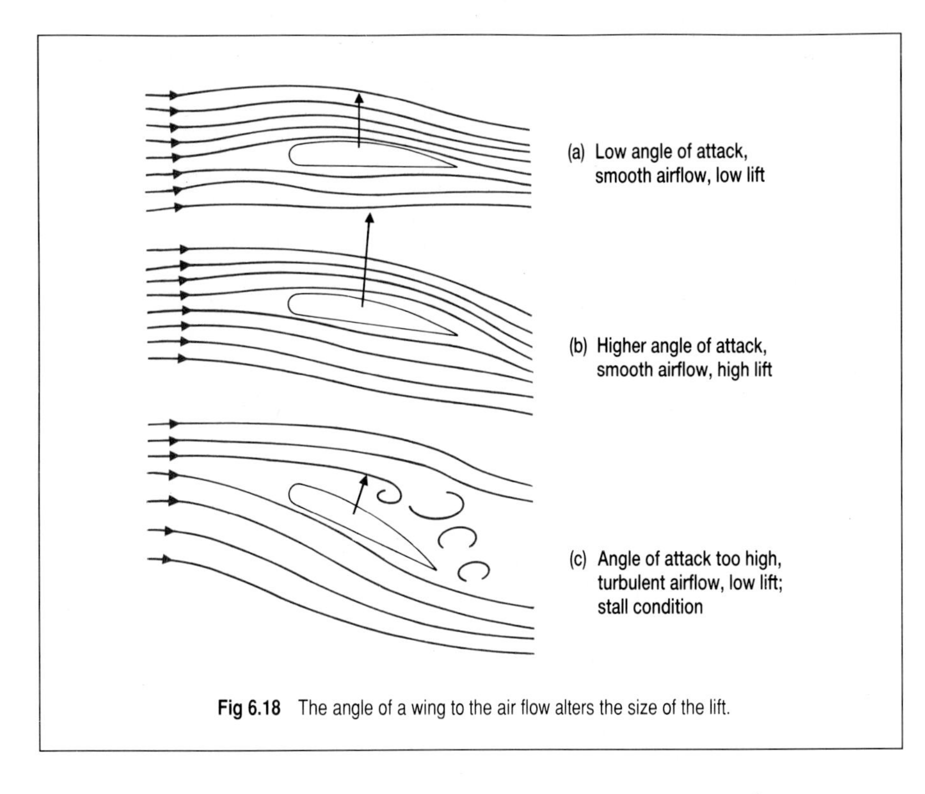

Fig 6.18 The angle of a wing to the air flow alters the size of the lift.

6.4 ARCHIMEDES' PRINCIPLE

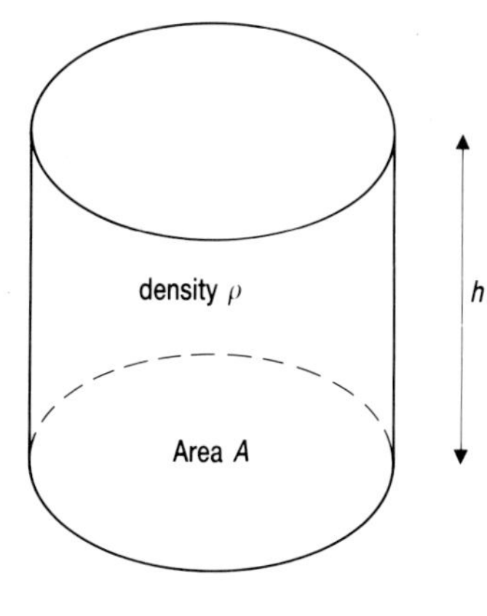

Fig 6.19

The fact that the pressure in a liquid increases with depth is well known. The equation relating pressure to depth is obtained as follows:

If a depth h of a column of liquid of density ρ is considered, then the mass of the column is $hA\rho$ where A is the area of cross section (Fig 6.19).

Weight of column = $hA\rho g$
so force on area A due to weight of liquid above it = $hA\rho g$
This gives pressure = force / area = $hA\rho g / A$
$= h\rho g$

This is the increase in the pressure caused by the liquid column of depth h. If the total pressure is required then the atmospheric pressure must be added on.

A submarine or other submersible vehicle must be strong enough to withstand the forces which are exerted on it by the surrounding sea. At a depth of 200 m in the North Sea, for example, the pressure due to the water of density 1030 kg m^{-3}, is given by

$$\text{pressure} = h\rho g = 200 \times 1030 \times 9.8 = 2\,020\,000 \text{ Pa}$$

or about 20 atmospheres. In the deepest trench in the Pacific Ocean the water is over 10 000 m deep and the pressure is over 1000 atmospheres. In 1960 a submersible vehicle called a bathyscape descended into this trench. It had a spherical compartment for the crew and was made out of steel 15 cm thick into which a wedge shaped window was fitted. The sphere was supported by a float which contained petrol, chosen because it was less dense than water. The float had openings at its bottom which allowed sea water to enter and squeeze the petrol. The pressure was high enough to reduce the volume of petrol to three-quarters of its original volume, see Fig 6.20.

One other important feature of the pressure in a fluid is that the forces it causes on any object act in a direction at right angles to the object's surface. A submerged object therefore has forces on it due to the surrounding water which are as shown in Fig 6. 21(a) and (b). The downward forces on the top of the object are less than the upward forces on the bottom of the object because the bottom of the object is in a greater depth of water than the top.

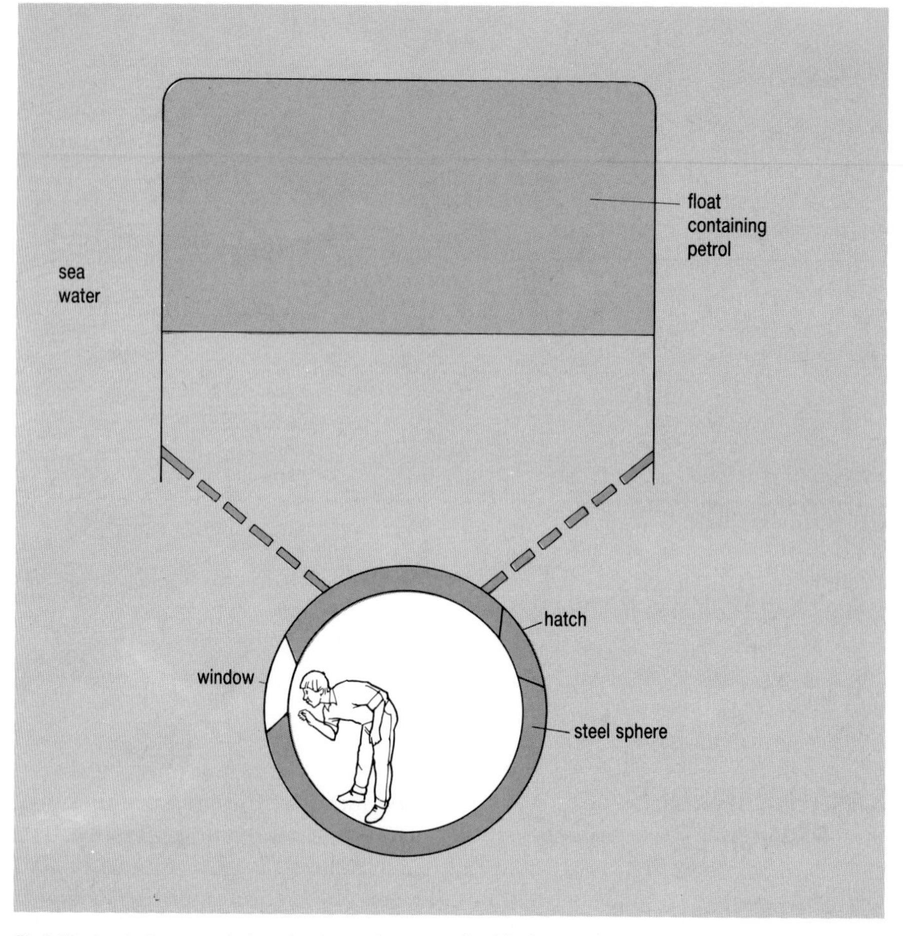

Fig 6.20 In a bathyscape, designed to descend to a great depth in the sea, the pressure of the water compresses the liquid in the float.

In still water sideways forces always cancel out no matter what the shape of the object. Since there is no resultant sideways force there must be a resultant upward force on the object. This force is called the **upthrust** or **buoyancy** force. Its value can be calculated for the cube in Fig 6.21(a).

pressure at top of cube $= h\rho g$
downward force on top = pressure × area $= h\rho g a^2$
pressure at bottom of cube $= (h + a)\rho g$
upward force on bottom $= (h + a)\rho g a^2$
upthrust $= (h + a)\rho g a^2 - h\rho g a^2$
$= h\rho g a^2 + a\rho g a^2 - h\rho g a^2$
$= a^3\rho g$

Note that the upthrust is not dependent on the depth to which the cube of Fig 6.21 is submerged (as long as it is under the surface) and is equal to the weight of a volume of liquid equal to the volume of the cube. The statement known as Archimedes' principle is a summary of this fact. It states:

> The upthrust on a body immersed in a fluid is equal to the weight of fluid displaced. The upthrust and the weight act in opposite directions.

If the body is floating then the principle can be extended to give:

> The weight of a body floating in equilibrium is equal to the weight of the fluid displaced.

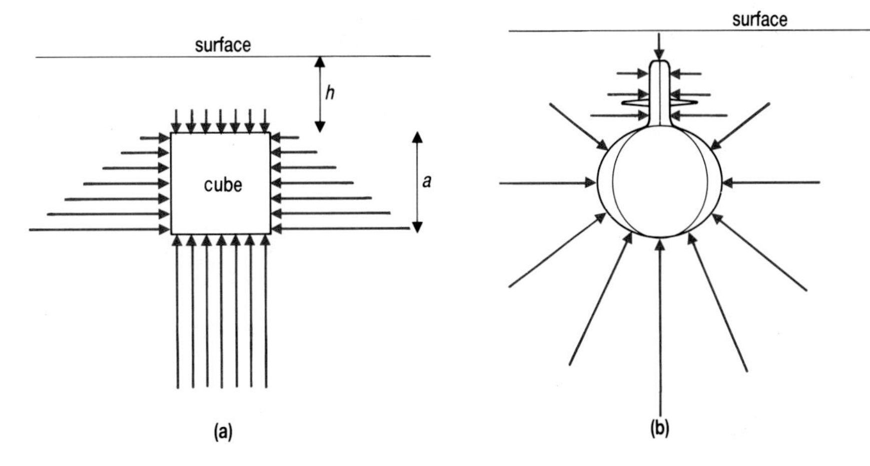

Fig 6.21 The forces exerted on a submerged object by the surrounding fluid shown for **(a)** a cube and **(b)** a submarine.

The **displacement tonnage** of large ships is given by marine engineers when they design a ship. The displacement tonnage is the sum of the ship's deadweight, which is all the cargo, fuels and stores, and its lightweight, which is the weight of hull, machinery and equipment. This figure is also equivalent to the mass of water which the ship displaces. As cargo is put aboard the ship it sinks lower in the water until the total weight of ship plus cargo is again equal to the weight of water displaced. A ship moving from fresh water into salt water will rise by a small amount as a smaller volume of sea water needs to be displaced for a given weight since the density of sea water is higher than that of fresh water.

QUESTION

6.8 The pressure of the atmosphere at sea level is the same as the pressure of a column of mercury 0.760 m high. The density of mercury is 13 600 kg m^{-3}. The density of air at standard pressure is 1.30 kg m^{-3}.

(a) What would be the height of the atmosphere if it had a constant density?

(b) State two factors which will alter the density of the atmosphere as the altitude increases.

(c) Explain why it is not sensible to give a definite value for the height of the atmosphere.

SUMMARY

The equation which controls all movement is

force	= mass	× acceleration
in newtons	in kilograms	in metre $second^{-2}$

In many forms of transport the force is provided by friction.
Resistance to the motion of a vehicle depends on its frontal area, the speed with which it is travelling, the materials surrounding it and its shape.
The Bernoulli effect is a reduction in the pressure exerted by a fluid when it flows.
Archimedes' principle can be used to find the upthrust on an object immersed in a fluid.

Chapter 7

EQUILIBRIUM

Any form of transport must be inherently safe. This means not only that it must be stable but also that it must be able to return to a stable state if, as will often be the case, it is displaced from a stable state whilst travelling. The important terms equilibrium and dynamic equilibrium will be discussed in this chapter.

LEARNING OBJECTIVES

After studying this chapter you should be able to:

1. know the conditions for the equilibrium of a body;
2. apply the principle of moments correctly to establish whether or not a body is in equilibrium;
3. find the power requirements of a moving vehicle;
4. understand the term dynamic equilibrium.

7.1 STATIC EQUILIBRIUM

Equilibrium on land

An object is said to be in equilibrium if it is not accelerating. The equilibrium of vehicles on land therefore must, at least initially, concern the vehicle when it is either stationary or moving with a constant velocity. To be in equilibrium when climbing a steep slope, a four wheel drive vehicle must have its weight acting in a line between the wheels, see Fig 7.1(a). If it is too heavily loaded at the back, as shown in Fig 7.1(b), it will be unable to

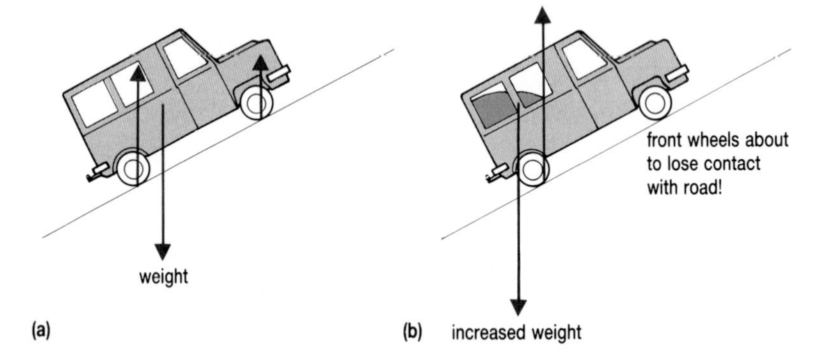

Fig 7.1 A four wheel drive vehicle on a steep slope. **(a)** in equilibrium, **(b)** not in equilibrium as a result of being overloaded at the back.

be in equilibrium and will topple backwards as a result of having a turning moment acting on it. The same is true of a bus on a steep camber. If the centre of gravity is too high, equilibrium is impossible and the bus topples over. A bus sustains equilibrium by having a massive chassis and hence a low centre of gravity even when passengers fill the upper deck (Fig 7.2).

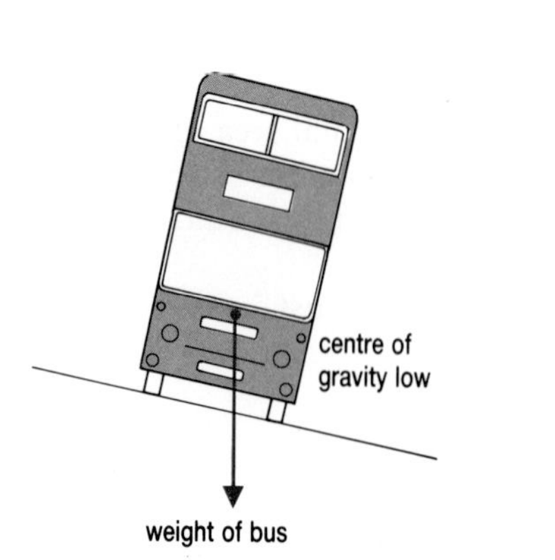

Fig 7.2 The low centre of gravity of a bus enables it to remain in equilibrium even when tilted to a large angle.

The problem of keeping the four wheels of a car on the ground when it is accelerating is more difficult. Clearly some condition is needed for this to occur but it cannot be one for equilibrium since the car is not in equilibrium if it is accelerating. A familiar experience for a speedway motorcyclist or

Fig 7.3 An easy stunt for a cyclist.

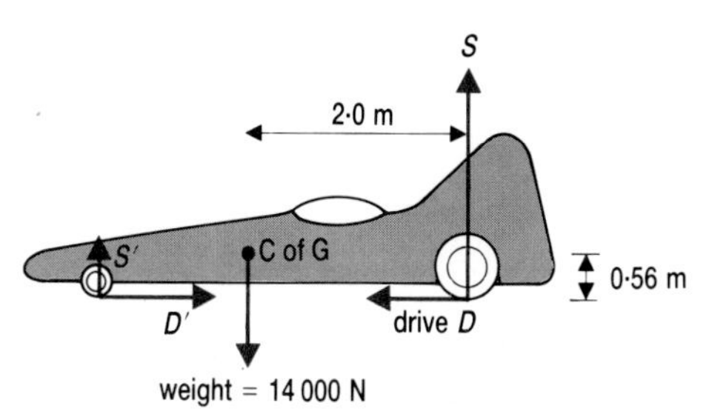

Fig 7.4 The forces on a dragster racing car.

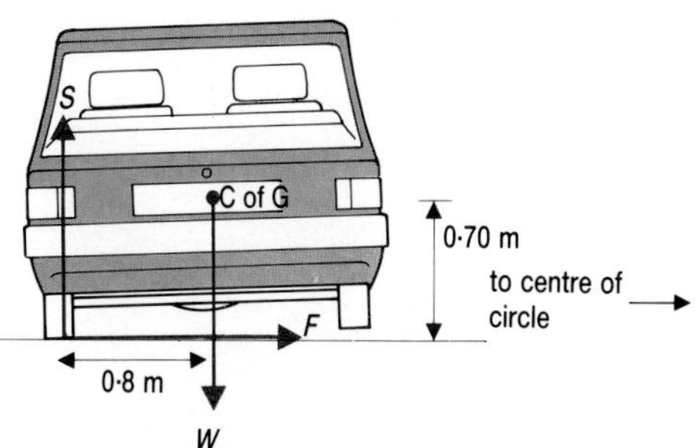

Fig 7.5 The forces acting on a car going round a corner. The car is seen from the back and is moving to the right.

perhaps for an ordinary motorcyclist is that by rapid acceleration the front wheel can be made to lift off the ground. Certainly it is not difficult for a pedal cyclist to make this happen (Fig 7.3). The mechanics of this situation for a car or cyclist depend on considering the acceleration and rotation of an extended object, a subject not covered in this book, but if moments are taken about the centre of gravity of the object then the correct result is obtained. (The correct result is *not* obtained if moments are taken about any point which is not the centre of gravity.)

One of the problems with dragster racing cars is how to avoid the front wheels lifting off the ground and so tipping the car over. Let us look at the maximum acceleration which the car can undergo without the front wheels lifting off the ground.

If the front wheels of the car in Fig 7.4 are just about to leave the ground then S′, the support force at the front wheels, and D′, the drag at the front wheels, are both zero.

By taking moments about the centre of gravity:
clockwise moments = anticlockwise moment if the car is not to rotate.

$$D \times 0.56 \text{ m} = S \times 2.0 \text{ m}$$

$$\text{i.e. } D = \frac{S \times 2.0 \text{ m}}{0.56 \text{ m}}$$

$$= 3.57S$$

In this circumstance the car is (just) not rotating so the vertical forces are equal and in opposite directions. S is therefore equal to 14 000 N.

$$\text{This makes } D = 3.57 \times 14\,000 \text{ N}$$

$$= 50\,000 \text{ N}$$

If there were sufficient frictional force between the back wheels and the ground this would cause the car, whose mass is 14 000/9.8 to accelerate.

$$\text{acceleration } a = \frac{\text{force}}{\text{mass}} = \frac{D}{m} = \frac{50\,000 \times 9.8}{14\,000}$$

$$= 35 \text{ m s}^{-2}$$

This question illustrates the importance of keeping the back wheels on the ground and explains why dragster cars have their characteristic shape. If the front wheels do come off the ground then the back wheels must support the entire weight of the car. This means that the maximum forward thrust is independent of the position of the centre of gravity. If its position is a long way forward, however, the car is less likely to come off the ground at the front.

For maximum D the centre of gravity needs to be low and as near the front of the car as possible. In this example an acceleration of 35 m s^{-2} requires a driving force of 50 000 N. Since the weight of the car is only 14 000 N a driving force of this magnitude is extremely unlikely. The tyres will slip on the road because the frictional force is insufficient. An acceleration of 35 m s^{-2} would enable a speed of 35 m s^{-1} to be reached one second after starting. This is over 70 mph. In practice, dragster cars can reach 60 mph in less than three seconds.

A similar problem of wheels coming off the ground during acceleration occurs when a car corners too rapidly. Here the acceleration is a centripetal acceleration and to study the forces we need to consider moments, which again are taken only about the centre of gravity. Let us look at a real example.

At what speed can a car round a corner of radius 40 m if the distance between its back wheels is 1.60 m and the centre of gravity is 0.70 m from the ground? (Fig 7.5). If the car is just able to take the bend then the inside wheels are just about to leave the ground. At this point $S = W$.

By moments about the centre of gravity:

$$S \times 0.80 \text{ m} = F \times 0.70 \text{ m}$$

so

$$F = \frac{0.80\ S}{0.70} = \frac{0.80}{0.70} \times W = \frac{8.0}{7.0} \times mg$$

since $W = mg$

where m is the mass of the car and g is the acceleration due to gravity.

So $$F = \frac{8.0 \times 9.8 \times m}{7.0}$$

Fig 7.6 Travelling round a corner at high speed can have disastrous consequences.

If the car is travelling at speed v and the radius of curvature is 40 m its centripetal acceleration is:

$$\frac{v^2}{r} = \frac{v^2}{40}$$

This acceleration is caused by the friction force F so since force = mass × acceleration

$$F = m \times \frac{v^2}{40}$$

$$11.2\ m = m \times \frac{v^2}{40}$$

$$11.2 \times 40 = v^2$$

$$v = \sqrt{11.2 \times 40} = 21 \text{ m s}^{-1}$$

Note that a higher cornering speed can be achieved if the wheels are more widely spaced and if the centre of gravity is as low as possible. This combination of requirements gives rise to the characteristic shape of a racing car. It also shows why it is that in stock car racing or even in rallying taking a corner too rapidly can lead to a somersault of the car.

Equilibrium of boats

If a boat, such as a canal barge, has an upthrust on it equal to its weight then it will be in equilibrium. The cross section of a canal barge is not very

Fig 7.7 When a narrow boat such as this crosses an aqueduct, no additional stress is put on the supports of the aqueduct. This is because the narrow boat displacesa weight of water equal to its own weight.

different from the shape of a box and so the boat will settle in the water until it displaces an amount of water equal in weight to its own total weight (Fig 7.7). If more cargo is placed on the boat then it will settle lower in the water until there is a sufficient increase in the buoyancy force to balance the increased weight of the barge and its cargo.

A barge on a canal is hardly representative of sea travel but all boats must be able to supply the necessary buoyancy force in the same way as a barge. At sea, however, they must be able to cope with adverse weather conditions causing rough seas. Since this involves what is called dynamic equilibrium it will be dealt with later in the chapter.

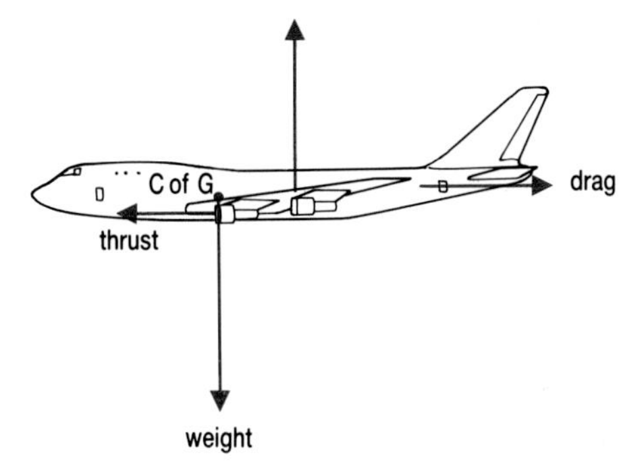

Fig 7.8 The forces acting on an aeroplane in flight.

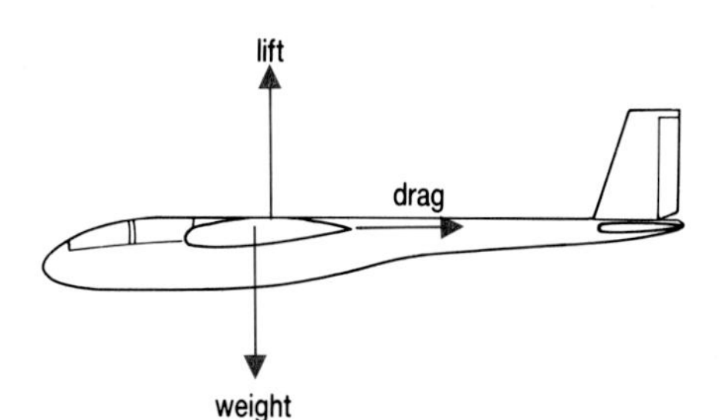

Fig 7.9 The forces acting on a glider in flight.

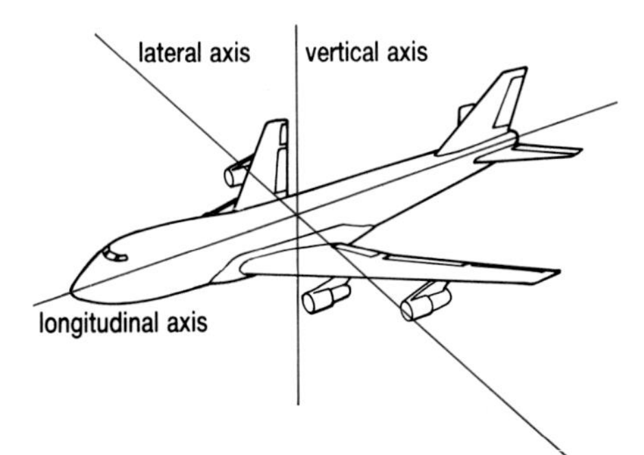

Fig 7.10 The three perpendicular axes referred to for an aeroplane all act through the centre of gravity of the aeroplane.

Equilibrium in the air

Section 6.3 mentioned the importance of the Bernoulli effect in supplying lift to an aeroplane. If the aeroplane is in level flight with a constant velocity, then the lift supplied to it by the Bernoulli effect must be equal and opposite to the weight of the aeroplane.

Lift is not the only effect which the air causes on an aeroplane. It also causes drag. To enable an aeroplane to travel with a constant velocity in level flight, therefore, not only must the weight equal the lift but a driving force, the **thrust**, must equal the drag. The thrust is provided by the engines. A typical diagram showing these four forces is given in Fig 7.8. Note that the lift does not necessarily act through the same point as the weight. In Fig 7.8 these two forces are shown acting in such a way that they will give an anticlockwise torque to the plane, making the nose go down. The thrust and drag, which are normally much smaller than the lift and weight, also provide a torque on the aeroplane. This is shown to give a clockwise torque. For straight and level flight, therefore, lift equals weight, thrust equals drag and any torque provided by lift and weight is balanced by an opposite torque provided by thrust and drag.

It is convenient, but an over-simplification, to suggest that there are just four forces acting on a plane. The lift acts all over the wings and the resultant of all these forces is taken to be a single force acting at a point called the **centre of pressure**. Similarly, drag acts at many points on the aeroplane and the resultant drag force acts through a point which depends on the design of the aeroplane. It is always considered as a force in a direction parallel to the airflow.

For a glider the diagram becomes Fig 7.9. A glider can only sustain a constant velocity if it is falling relative to the air surrounding it.

So far aeroplanes have only been considered in level flight at a constant velocity. Obviously, aeroplanes need to be able to change their velocity and they can do this by increasing or decreasing the thrust of the engines or by the use of various control surfaces.

Being a three-dimensional object, an aeroplane may change the direction of its velocity by rotating about three perpendicular axes (Fig 7.10). If it lifts its nose it is rotating about the lateral axis; if it changes the direction in which it is travelling it is rotating about its vertical axis; and if it dips one wing and raises the other then it is rotating about its longitudinal axis. Stability in the air necessitates control of all these rotations and since each concerns dynamic equilibrium they will be considered later in the chapter.

QUESTIONS

7.1 A car ferry across a lake can carry 24 loaded cars/vans/minibuses of average mass 1600 kg. The area of the ferry at the water line is 150 m^2 and this area may be considered a constant as the boat settles lower in the water. By how much does the ferry settle as the cars are loaded?

7.2 **(a)** A modern jet airliner, travelling with a velocity of 250 m s^{-1}, has wings of total area 360 m^2 and has a total mass of 100 000 kg. What is the average pressure difference, due to the Bernoulli effect between the top and bottom surfaces of the wings at this speed?

(b) What is this pressure difference as a fraction of atmospheric pressure, 40 000 Pa, at the height at which the flight takes place?

7.3 The weight of an aeroplane, acting through its centre of gravity, is 700 000 N. The lift acts through a point 0.83 m behind the centre of gravity. The total drag on the aircraft when it is flying with constant velocity in level flight is a force of 75 000 N.
Show the forces acting on the aeroplane on a diagram and calculate:
(a) the lift;
(b) the thrust;
(c) the turning effect of the lift and weight;
(d) the turning effect of the thrust and drag;
(e) the vertical distance necessary between the lines of action of the drag and thrust.
Comment on your answer to part (e) and suggest where the engines may be positioned.

7.2 POWER REQUIREMENTS

Power needed when in equilibrium

To calculate the power required by an aeroplane when it is in straight, level flight with a constant velocity is straightforward.

Consider an aeroplane providing a thrust of 80 000 N travelling in straight level flight at a speed of 250 m s^{-1}.

The work done by the engines in one second is:

$$\text{force} \times \text{distance moved in direction of force} = 80\,000\ \text{N} \times 250\ \text{m} = 20\,000\,000\ \text{J}$$

Since this is the work done in one second the power is 20 000 000 W or 2×10^7 W. The power is given by:

$$\text{power} = \text{thrust} \times \text{velocity}$$

This equation holds whether the aeroplane is accelerating or not. It is a general equation which can be applied to all vehicles, not just aeroplanes.

Power needed to accelerate

A problem arises when it is the power which is fixed rather than the velocity. If a car starts from rest and can give a power output of 80 kW, then when it is travelling slowly the force of traction can be very large. At 2 m s^{-1}, for example, the horizontal force exerted by the wheels on the road F could be given by:

$$F \times 2\ \text{m s}^{-1} = 80\,000\ \text{W}$$

$$F = 40\,000\ \text{N}$$

This force is much larger than the likely value of the frictional force so the drive wheels skid on the road surface. This is a common occurrence for a driver who tries to get away from traffic lights too quickly.

At the other end of the scale the equation limits the maximum speed of the car. If total frictional and drag forces on the car are 2000 N at 40 m s^{-1} then the power output is correctly 2000 N × 40 m s^{-1} = 80 000 W. The car is unable to accelerate further because a higher speed would imply an even greater frictional force and this would only be possible, from the point of view of power, if the velocity were lower.

QUESTION

7.4 A car has a maximum output power of 140 kW, a mass of 1200 kg, a maximum frictional force between the driving wheels and the road which is one quarter of its weight and a total drag force D, on it given by the equation:

$$D = k + cv^2$$

where k is a constant with the value 400 N, c is a constant with the value 1.00 kg m^{-1} and v is its velocity. Find:

(a) the maximum acceleration of the car;
(b) the maximum speed of the car;
(c) the speed the car would be able to travel at if not limited by the consideration of power output;
(d) the maximum deceleration of the car.

Draw a sketch graph to show:

(e) how the speed of the car varies with time during its acceleration;
(f) how the speed–time graph differs for a car not limited by power output.

7.3 DYNAMIC EQUILIBRIUM

Equilibrium of several vehicles has been mentioned already. It has been stated that a body is in equilibrium if it is not accelerating. However, for many objects, including vehicles, static equilibrium is insufficient for safety. This does not only apply to modes of transport. A bridge, for example, must be in equilibrium but it must also be in dynamic equilibrium. That is, if it is for some reason moved from its rest position it must return to its rest position as soon as the reason for its movement is removed. Bridges do move as a result of wind acting on them. If the wind sets up ever increasing oscillations of the bridge then it is not in dynamic equilibrium. This was the problem with the Tacoma Narrows bridge in the State of Washington, USA in 1940 (Fig 7.11).

Fig 7.11 As a result of oscillations set up by the wind, this spectacular collapse of the Tacoma Narrows bridge took place in 1940.

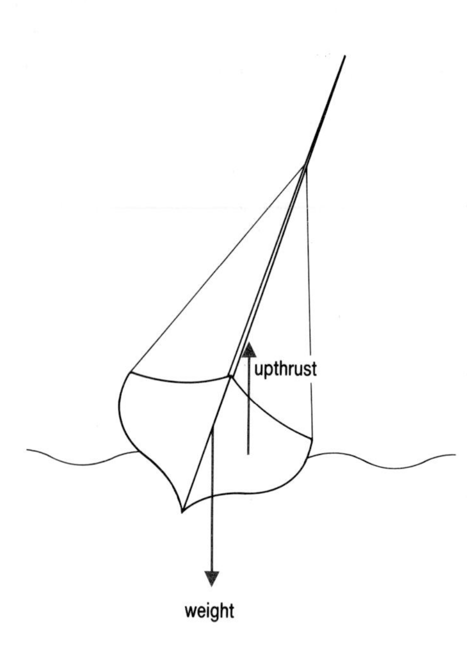

Fig 7.12 When any boat is displaced from equilibrium by the sea, forces must come into effect which return the boat to equilibrium if the boat is not to capsize.

Dynamic equilibrium concerns the way in which an object regains its original position after it has been displaced slightly. A boat in equilibrium is bound to be displaced from that equilibrium by a wave. It must regain equilibrium and not capsize. All boats are designed so that they regain equilibrium from up to a certain angle of rotation. Some lifeboats are designed so that they will regain their original position even if rotated through 180°, though this is usually achieved by special methods involving internal water tanks rather than by simple considerations of equilibrium. Normally equilibrium is regained by arranging that a boat which rotates clockwise has a restoring anticlockwise couple acting on it, as shown in Fig 7.12.

The weight of the boat acts through its centre of gravity and the upthrust acts through a point called the **centre of pressure**. This point is the centre of gravity of the displaced water. In order to get a large restoring couple many boats bulge outwards at the side so that a large volume of water is displaced as far from the centre of gravity as possible.

There is another consideration besides the provision of a restoring couple and this concerns the time taken to return to equilibrium and the period of oscillation. A well designed boat will not swing backwards and forwards frequently if displaced once. The keel helps to reduce oscillation: it cannot stop it. The damping of the oscillations needs to be carefully controlled. Too much damping means that the system stays too long in the displaced position so that further buffeting may occur before it has regained equilibrium. Critical damping is shown in Fig 7.13. This amount of damping gives the shortest time to equilibrium without any oscillation occurring.

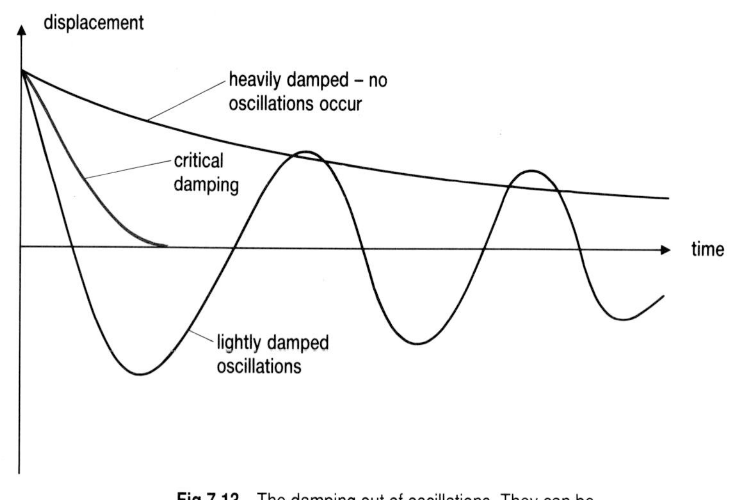

Fig 7.13 The damping out of oscillations. They can be heavily or lightly damped.

INVESTIGATION

Damping oscillators

Find how the shape of a floating piece of wood affects its stability.
Use a series of pieces of softwood of different width-to-depth ratios and float each on water (Fig 7.14). The length of the wood is not important, provided it is longer than either the width or the depth.

- Which pieces of wood float in equilibrium? Which are also in dynamic equilibrium?
- For what range of width-to-depth ratios is the top surface of the wood horizontal?
- How does the damping of the oscillations depend on the width to depth ratio?
- How could the stability of the unstable blocks be improved?

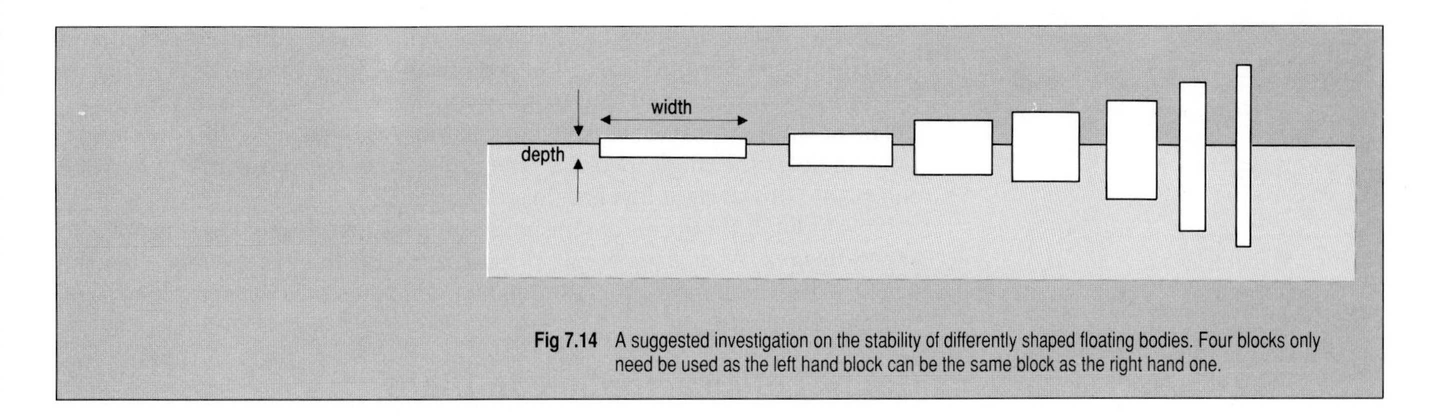

Fig 7.14 A suggested investigation on the stability of differently shaped floating bodies. Four blocks only need be used as the left hand block can be the same block as the right hand one.

Dynamic equilibrium of aeroplanes

The same dynamic equilibrium problem arises in the design of an aeroplane. In order to have dynamic stability an aeroplane must return to its equilibrium position if it is displaced from it by air turbulence. As stated earlier in the chapter, an aeroplane may rotate about one, two or three axes. Now let us consider each separately.

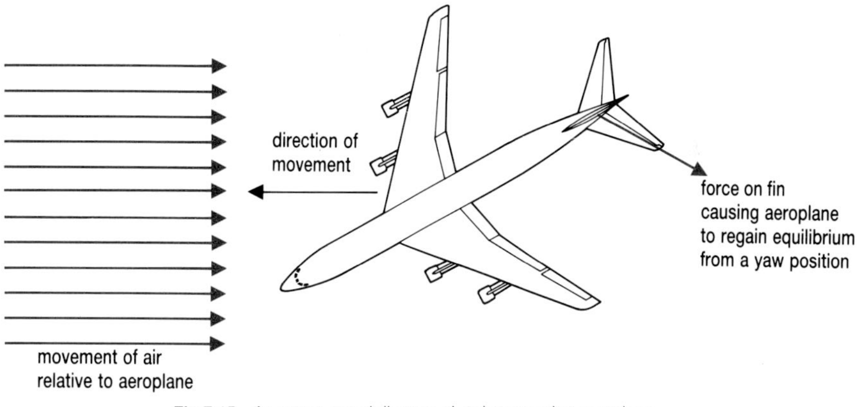

Fig 7.15 An exaggerated diagram showing a yawing aeroplane.

Fig 7.15 shows, in an exaggerated way, an aeroplane which has rotated about its vertical axis. The plane is said to have **yawed**. As a result of this the fin of the aeroplane is partially sideways on to the airstream and so there is a force on the fin which causes a turning moment on the aeroplane to rotate it in a clockwise direction. Only when the fin is parallel to the airflow is there no turning moment. This area of the fin can be designed to give sufficient force for critical damping.

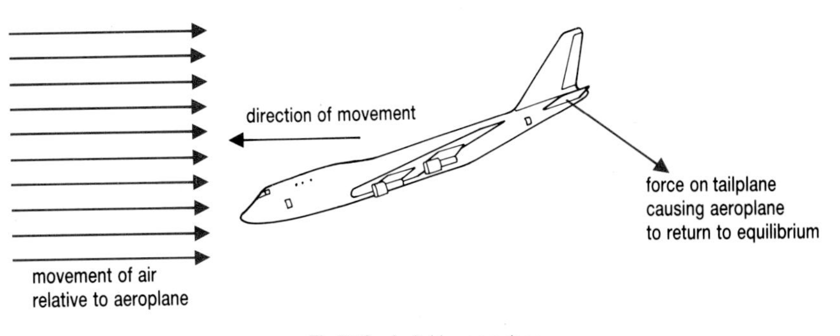

Fig 7.16 A pitching aeroplane.

Fig 7.16 shows an aeroplane which has been rotated about its lateral axis. The plane is said to have **pitched**. Here the tailplane is travelling partially sideways through the air and so the force in this case which causes a restoring moment is due to the tailplane.

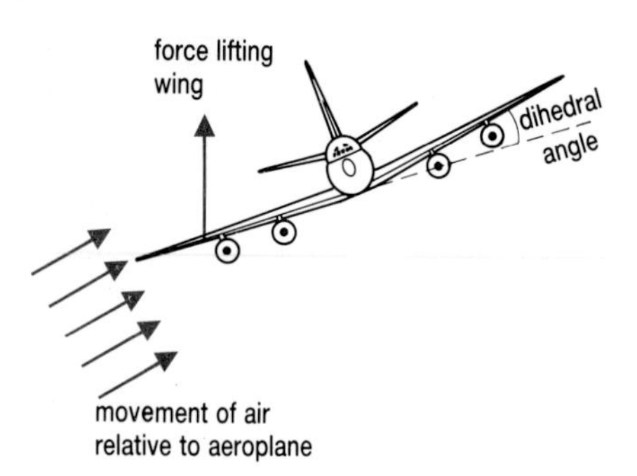

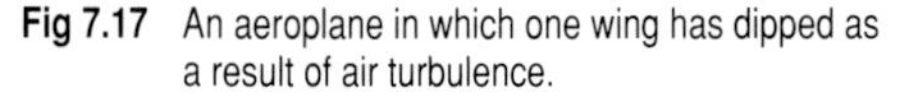
Fig 7.17 An aeroplane in which one wing has dipped as a result of air turbulence.

Rotation about the longitudinal axis is rather more subtly corrected and in fact different designers do allow for this in different ways, depending on the type of aeroplane and the speeds with which it is likely to travel.

The most usual way of establishing stability is by using a **dihedral angle**. This angle is the angle between the wings and the horizontal when the aeroplane is in straight level flight.

If, as in Fig 7.17, the aeroplane rolls about its longitudinal axis it starts to slip sideways. As it does so air presses more on the lower wing than the upper wing and a moment is established which restores the aeroplane to its normal flying attitude.

Control of an aeroplane in flight

The dynamic equilibrium of an aeroplane in flight is ensured by its designer. The aeroplane must be designed to be inherently stable. During a flight, however, the equilibrium of the aeroplane can be disturbed by the pilot who needs to be able to change the direction in which the aeroplane is flying. This he does by using control surfaces which are attached to wings, tailplane and fin.

These are shown in Fig 7.18 which also shows the forces exerted on the control surface. In Fig 7.18(a) elevators are shown in a down position. The extra force on the elevators causes a moment rotating the aircraft about its lateral axis so that it will point nose down. If the elevators are moved to an up position then there is a clockwise moment on the aircraft and it goes nose up.

In Fig 7.18(b) the force on the rudder causes rotation of the aeroplane about its vertical axis. This enables a pilot to point the plane in a chosen direction. In practice, if a rudder manoeuvre only is carried out the aeroplane tends to yaw, crab-like. It 'continues to move in its line of motion in a straight line' to quote Newton's First Law, rather than to change the direction of travel.

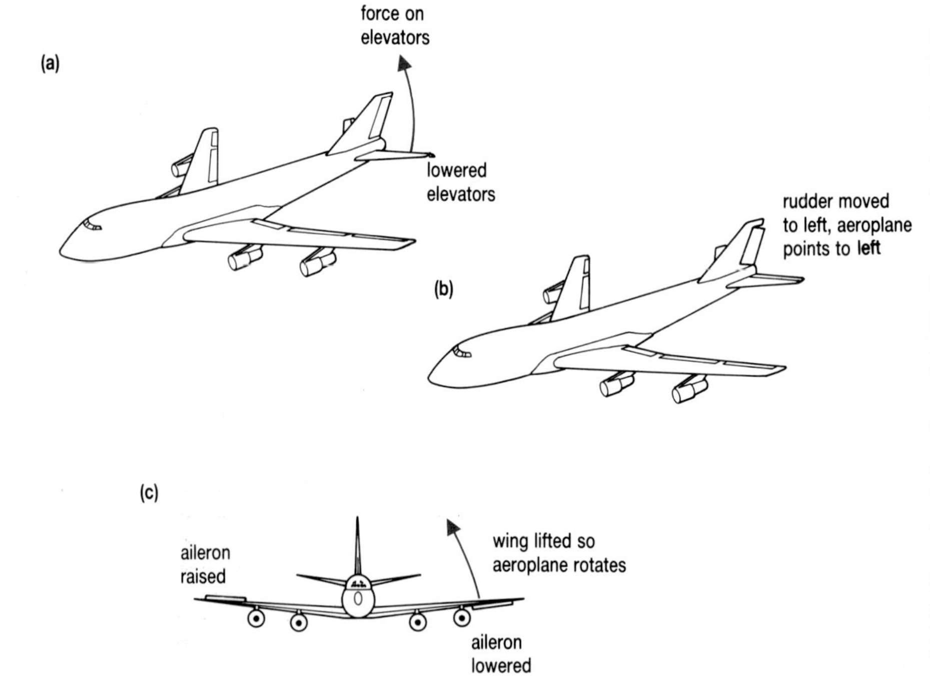

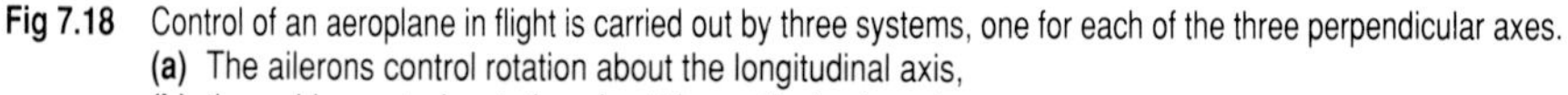
Fig 7.18 Control of an aeroplane in flight is carried out by three systems, one for each of the three perpendicular axes.
(a) The ailerons control rotation about the longitudinal axis,
(b) the rudder controls rotation about the vertical axis and
(c) the elevator controls rotation about the lateral axis.

To change course a pilot will use the rudder but will also use the ailerons to bank the aeroplane, that is, to rotate it about its longitudinal axis. The ailerons are hinged flaps attached to the trailing edges of the outer parts of the wings. They are moved in opposite directions. In Fig 7.18(c) the right aileron has been moved downwards and the left one upwards. This gives the right wing extra upward force and the left wing less than normal upward force. The overall effect is to provide the aeroplane with an anticlockwise moment which rotates it about its longitudinal axis.

SUMMARY

An object is in equilibrium if the resultant force on it is zero. All forms of transport are in equilibrium when travelling in a straight line at constant speed.

An aeroplane's stability illustrates dynamic equilibrium well. If the plane meets turbulent air it will be disturbed from its equilibrium state. Its design must be such that it will regain equilibrium promptly. The fact that it regains its equilibrium position implies that it must be in dynamic equilibrium when flying.

Examination Questions

T3.1

(a) Explain how the forward propulsive force on an aircraft is provided by:
   (i) a propeller
   (ii) a jet engine

(b) An aircraft is in straight, level flight at a constant speed of 200 m s^{-1}. At a certain time, the mass of the aircraft and its fuel is 4×10^8 kg and fuel is being consumed at the rate of 3 kg s^{-1}. The chemical energy stored in the fuel is 45 MJ kg^{-1}.
   (i) What is the kinetic energy of the aircraft at this instant?
   (ii) Assuming that fuel continues to be consumed at this rate and no other adjustments are made, state and explain what will happen to the speed and the height of the aircraft.
   (iii) What is the rate of energy production from the combustion of the fuel?
   (iv) Assuming that the overall efficiency of the aircraft's engine is 25%, estimate the maximum drag experienced.

(UCLES specimen, 1990)

T3.2

(a) Explain briefly what is meant by a *drag force*.

(b) A stationary plane surface of area A is held at right angles to a stream of air moving at speed v. Show that F, the force exerted on the surface by the air stream, is given by $F = kAv^2$, where k is a constant.

(c) In a certain vehicle travelling at 100 km h^{-1}, the power required to compensate for drag is 45 kW.
   Calculate
   (i) the drag force at this speed,
   (ii) the area of the equivalent plane surface described in **(b)** which would produce the same drag, given that $k = 1.5$ kg m^{-3}.

(d) When towing a caravan, an inclined plane is sometimes fitted to the roof of the towing vehicle, as shown in the diagram.
   Give a physical explanation of its purpose.

(UCLES, 1990)

T3.3

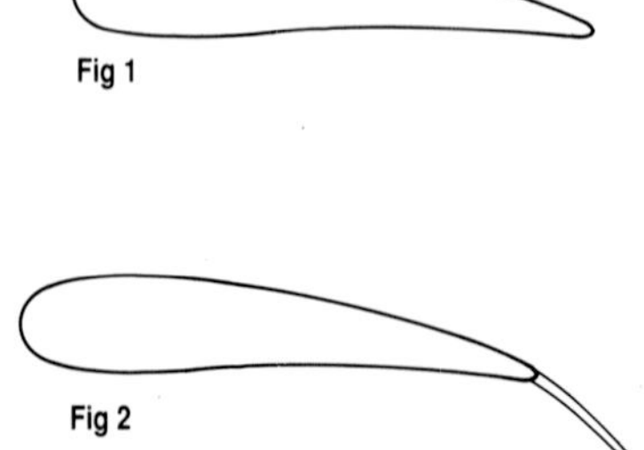
Fig 1

Fig 2

(a) What do you understand by the *Bernoulli effect*?

(b) (i) An aircraft is cruising at a constant altitude. Fig 1 illustrates a cross-section through the wing. Explain how a lift force is generated.
   (ii) When coming in to land, the trailing edge of the wing may be extended, as shown in Fig. 2. Explain why this increases the lift.

(c) It is well known that some aircraft can sustain level flight whilst upside down. Suggest an explanation of how this is possible.

(UCLES, 1990)

T3.4

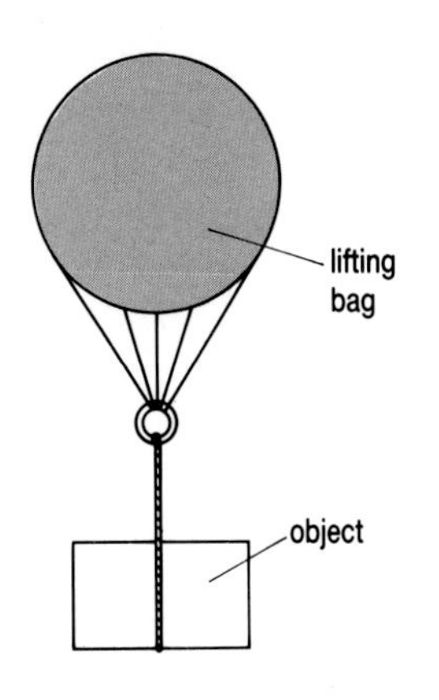

(a) Steel has a greater density than water and yet a steel ship can float in water. Explain the physical principles involved, including the achievement of stability.

(b) In order to raise a heavy object from the sea-bed, a 'lifting bag' may be attached to the object and then partially inflated with air, as shown in the diagram. Explain why air has to be released continuously from the lifting bag as the object rises to the surface so that a constant speed of ascent is maintained.

(c) A submerged iron cannon of mass 800 kg and density 8000 kg m^{-3} is attached to a lifting bag of negligible volume and mass. Estimate the initial acceleration of the cannon when 0.70 m^3 of air is suddenly released into the bag. (The density of the water is 1050 kg m^{-3}.)

(UCLES, 1990)

T3.5

(a) A car engine raises the temperature of the fuel to about 600 °C and expels combustion products at about 80 °C. Calculate the efficiency of the engine if it achieves 70% of the maximum theoretical efficiency of a 'perfect' heat engine working between these temperatures.

(b) The diagram shows the engine efficiency as a function of car speed, for a car in top gear on a level road. Describe the variation of frictional losses and pumping losses with speed. Describe the causes of frictional losses and pumping losses in the engine.

(ULSEB, 1990)

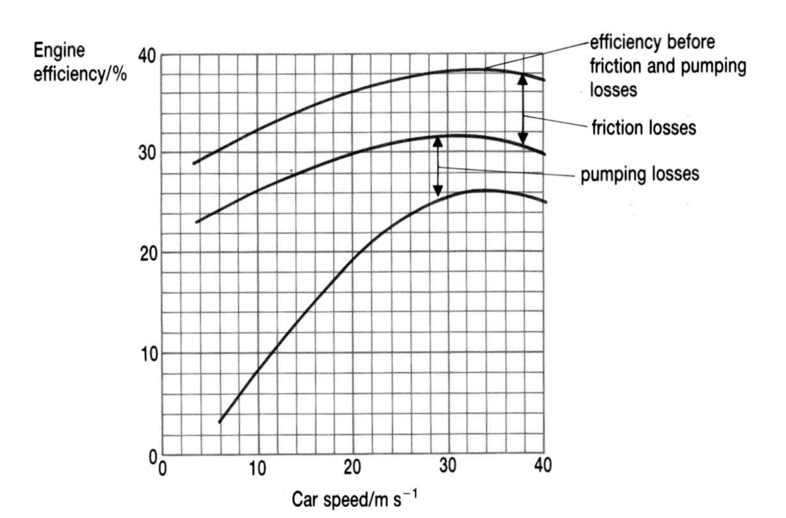

Appendix A
ANSWERS TO QUESTIONS

Chapter 1

1.2 40 m approx.
1.3 2.9×10^{-11} J
1.4 2.8×10^{-12} J

Chapter 2

2.1 **(a)** Demand is spread
(b) Cheaper
(c) Increased demand, early mornings in winter
(d) Gas can be stored easily
2.2 Approximately £100 extra

Chapter 3

3.2 Largest: space and water heating; smallest: lighting etc.
3.4 15%
3.6 6.5% approx.
3.7 Coal: 4060 million tonnes
Oil: 2940 million tonnes
Gas: 1850×10^9 m^3
Hydroelectric: 6.6×10^{12} kW h
Nuclear: 5.1×10^{12} kW h
3.8 **(a)** 1.4×10^{11} **(b)** 5.4 tce

Chapter 4

4.1 1940 MW
52.4 kg s^{-1}
4.2 **(a)** 44 W
(b) twice the power loss so heater on for 20 s every $7\frac{1}{2}$ min.
4.3 **(a)** 6.0×10^{-21} J
(b) 480 m s^{-1}
(c) 510 m s^{-1}
4.4 a, d, e
4.5 **(a)** 200 000 J
(b) 3×10^{26} J
4.6 Internal energy decreases by 1000 J so temperature of gas falls
4.7 800 J
500 K

Chapter 5

5.1 **(a)** 0.125
(c) Absolute zero
5.2 **(a)** 0.55
(b) 0.68
5.4 **(a)** 3300 MJ
(b) 2100 J
5.5

1.95p	48p
3.0p	75p
7.0p	175p
3.5p	88p

5.6 **(a)** 3 cm^3 s^{-1}
(b) 9.0 litres
5.7 **(a)** 1300 W
(b) It falls and becomes zero at absolute zero

Chapter 6

6.1 **(a)** 320 N
(b) 180 N
(c) 0.80 m s^{-2}
(d) 720 N
(e) 1420 N
(f) 1040 N
6.3 0.5 m s^{-2}, 2.0×10^5 N, 3600 m
6.4 1500 N, 0.50 m s^{-2}
6.5 5.2 m
6.6 2.3 m
6.7 240 000 N
6.8 **(a)** 8000 m

Chapter 7

7.1 26 cm
7.2 **(a)** 2720 Pa
(b) 0.068
7.3 **(a)** 700 000 N
(b) 75 000 N
(c) 581 000 N m
(d) 581 000 N m
(e) 7.75 m
7.4 **(a)** 2.2 m s^{-2}
(b) 46.7 m s^{-1}
(c) 50.4 m s^{-1}
(d) 4.6 m s^{-2}

Examination questions

T1.1 **(c)** 62.6 ms^{-1}; 217 MW; 173 MW
T1.2 Available energy = 4.1×10^{21} J
T1.3 **(a)** (i) 1882 kW h (ii) 4788 kW h (iii) £133 (iv) £208
(b) 6.66×10^9 J
T1.4 **(a)** (i) 8.86×12 kg (ii) 21 years
T2.1 **(b)** (i) 6.8×10^6 J (ii) ~0.2 kg
(c) 210 W
T2.2 **(c)** Energy requirement ~40 MJ
T2.3 **(a)** (i) 12 Kh^{-1} kW^{-1} (ii) 1.5 kg h^{-1} kW^{-1}
(b) (i) 8 K (iii) 0.45 kg (iv) 3.2 kg, dangerously near 5% of body mass.
T3.1 **(b)** (i) 8×10^{12} J (ii) Drag unaltered by reduced mass. Thrust unaltered therefore speed remains constant. Weight reduced, lift unaltered therefore aircraft rises. (iii) 135 MW (iv) 1.7×10^5 N
T3.2 **(c)** (i) 1600 N (ii) 1.4 m s^2
T3.4 **(c)** 0.49 m s^2
T3.5 **(a)** 42%

Appendix B
GUIDELINES FOR THE ASSIGNMENTS

Chapter 1

Making use of solar cells

1. In space technology, the cost of individual components is often negligible compared to total cost of launching a space vehicle.
2. You may assume that the solar panels can be oriented to make best use of the Sun's radiation.
3. Use the inverse square law.
4. The higher the latitude, the greater the area needed, as the solar cells are tipped away from the direct rays of the sun. This introduces a factor of $1/\cos\theta$.

The Foyers hydroelectric scheme

1. Use $E_p = mgh$.
2. Use mgh $= \frac{1}{2}mv^2$.
3. Overall efficiency is high, over 80%.
4. Base your calculation on the mass of water which passes through in 1 second.
5. Of course, the wasted energy is likely to be transformed eventually to heat, but you need to suggest ways in which this happens.

France's tidal power station

1. You may assume that the basin is fairly flat-bottomed, and so the volume of water = area × depth.
3. The text tells you the power output of the system; you will have to take account of the fraction of the time during which the station can operate.
4. One of the problems of tidal power stations is that they cannot operate at a uniform level of output, because of the nature of tides, and the variation of the times of high tide from day to day.

Energy from waves

1. The duck extracts energy from the waves, and so their amplitude must be reduced.
4. It may help to think about how alternating electric current can be rectified using diodes. A simple flap valve will allow air to pass through in one direction only.

The Furlmatic wind generator

4. The power output of a wind generator is usually quoted at a level below its maximum possible output, to give a more realistic picture of its potential.
5. Consider the upward-sloping part of the graph. Take three or four values of power and wind speed from the graph, and test them for cubic variation.

Energy from fuels

Since hydrogen contains no carbon, it must be top of your league table!

Water power

1. First, calculate the number of hydrogen atoms in 1 kg of water (2 per molecule). Then find the number which are deuterium. Remember that two of these are needed for each fusion event. How much energy can they provide?

2. Of course, we are assuming we could carry out the process with 100% efficiency.
3. How does this answer compare with the age of the Earth, $4\frac{1}{2}$ billion years?

Hot rocks energy
1. You need to understand the term 'specific heat capacity'.
3. Because of the relatively low temperature of the hot water, such a system is likely to have low efficiency.

Energy origins and transfers
1,2. Energy from recent solar radiation is usually regarded as renewable. You may find uranium referred to as a renewable resource; this is because in a breeder reactor it can be used to make plutonium fuel.
3. Block diagrams are a simple way of showing energy transformations.

Chapter 2

Transformer efficiencies
2. The largest transformers are the most efficient. This may be because large machines are generally more efficient than small ones; it may also reflect the fact that more research effort has been put into making these large machines more efficient.

Variations in demand
1. You should notice that the lowest summer demand is only a small fraction (perhaps one quarter) of the highest winter demand.
2. Consequently, the number of stations required to be operating varies greatly throughout the year.

Living with building regulations
1. Calculate the rate of heat loss for window and wall separately, and then add them. Which is the greater of the two?
2. Why might the windows not be double-glazed?

Chapter 3

Energy trends
1,2. Think about the historical development of the UK, and of the USA.
3. Think about the availability of suitable water sources, and of the necessary technology. What technology might be developed to allow further use of hydroelectric power?

Living in a greenhouse
Information about global climate changes is accumulating rapidly, and evaluations of future prospects change correspondingly. You should be able to find some up-to-date articles which will give you some idea of current opinions. Try to establish whether there are differing views, and what evidence there is to support these.

Coping with nuclear waste
When reviewing evidence from a variety of sources, you should be able to identify clear differences of opinion. Try to think about the different assumptions which underlie these differences of view.

Chapter 4

Heat engine efficiencies

Heat engines generally have the lowest efficiencies, below 50%.

Chapter 5

Data analysis

1. 0.044 mol
2. Since compression ratio is 16 : 1, volume is one-sixteenth of original volume.
7. Work done by gas = 850 J; Heat supplied to gas = 1400 J; Efficiency = 0.61.
8. 690 J
9. 16 g s^{-1}, 0.067 g per injection
10. 1400 J
11. 2.1×10^6 J kg^{-1}

Chapter 6

Cars and air resistance

1. 550 N
2. 700 N
3. 700 N

Appendix C

FURTHER RESOURCES

Books and other resources

1. **A resource directory**
   Energy across the curriculum – a resource directory for GCSE. Department of Energy, 1987. (Although this directory is intended to relate to syllabuses at GCSE level, many of the resources listed are suitable for more advanced students.)

2. **Other textbooks**
   There are many books which deal with particular aspects of energy and transport. Those listed here give a broader coverage of these subjects.
   Energy: A guidebook, by Janet Ramage. Oxford, 1983.
   Energy resources, by J T McMullan, R Morgan and R B Murray. Edward Arnold, 1983.
   Energy through time, by Joe Scott. Oxford, 1986.
   Energy without end, by Michael Flood. Friends of the Earth, 1986.
   Fuel and energy, by H Backhurst. Academic Press, 1981.
   Physics: Energy options, ed. Maurice Tebbutt, Nuffield Advanced Physics. Longman, 1986.
   Solar prospects, by Michael Flood. Wildwood House, 1983.

3. **SATIS materials**
   The Science and Technology in Society (SATIS) project has published many useful teaching resources concerned with energy and transport. A guide to units dealing with aspects of energy is published in the *General Guide for Teachers,* in the SATIS 16–19 series.

   SATIS is published by the Association for Science Education.

Useful addresses

The organisations listed below have a broad interest in many aspects of energy supply and use. They provide much useful resource material, suitable for use in schools and colleges. Much of this is available free; lists are usually available on request. Some of the organisations also supply videos and speakers.

In addition, many more addresses may be found in *Energy across the Curriculum,* as mentioned in **Books and other resources**. These include organisations with interests in specific aspects of fuel production, and energy supply and consumption.

Centre for Alternative Technology, Machynlieth, N. Wales.

Centre for Research, Education and Training in Energy (CREATE), 18 Devonshire Street, London W1N 2AU.

Energy Technology Support Unit (ETSU), Building 156, Harwell Laboratory, Didcot, Oxfordshire OX11 0RA.

Energy Efficiency Office, Department of Energy, Thames House South, Millbank, London SW1P 4QJ.

Friends of the Earth Trust Ltd, 26–28 Underwood Street, London N1 7JQ.

Index

Bold page references denote key entries.